EFFICIENT MANAGEMENT OF WASTEWATER FROM MANUFACTURING

MANUFACTURING

New Treatment Technologies

EFFICIENT MANAGEMENT OF WASTEWATER FROM MANUFACTURING

New Treatment Technologies

Edited by
Victor Monsalvo, PhD

Apple Academic Press Inc. | Apple Academic Press Inc.
3333 Mistwell Crescent | 9 Spinnaker Way
Oakville, ON L6L 0A2 | Waretown, NJ 08758
Canada | USA

©2015 by Apple Academic Press, Inc.

First issued in paperback 2021

Exclusive worldwide distribution by CRC Press, a member of Taylor & Francis Group
No claim to original U.S. Government works

ISBN 13: 978-1-77463-568-1 (pbk)
ISBN 13: 978-1-77188-171-5 (hbk)

Library and Archives Canada Cataloguing in Publication

Efficient management of wastewater from manufacturing: new treatment technologies / edited by Victor Monsalvo, PhD.

Includes bibliographical references and index.
ISBN 978-1-77188-171-5 (bound)
1. Factory and trade waste--Purification--Technological innovations. 2. Sewage--Purification--Technological innovations.
I. Monsalvo, Victor (Victor M.), author, editor

TD897.E34 2015   628.1'683   C2015-900729-1

Library of Congress Cataloging-in-Publication Data

Efficient management of wastewater from manufacturing: new treatment technologies / editor, Victor Monsalvo, PhD.

pages cm
Includes bibliographical references and index.
ISBN 978-1-77188-171-5 (alk. paper)
1. Sewage--Purification. 2. Factory and trade waste--Management. I. Monsalvo, Victor M.

TD745.E434 2015   628.1'683--dc23   2015002644

Apple Academic Press also publishes its books in a variety of electronic formats. Some content that appears in print may not be available in electronic format. For information about Apple Academic Press products, visit our website at **www.appleacademicpress.com** and the CRC Press website at **www.crc-press.com**

About the Editor

VICTOR MONSALVO, PhD

Professor Victor Monsalvo is an environmental scientist with a PhD in chemical engineering from the University Autonoma de Madrid, where he later became a professor in the chemical engineering section. As a researcher, he has worked with the following universities: Leeds, Cranfield, Sydney, and Aachen. He took part of an active research team working in areas of environmental technologies, water recycling, and advanced water treatment systems. He has been involved in sixteen research projects sponsored by various entities. He has led nine research projects with private companies and an R&D national project, coauthored two patents (national and international) and a book, edited two books, and written around fifty journal and referred conference papers. He has given two key notes in international conferences and has been a member of the organizing committee of five national and international conferences, workshops, and summer schools. He is currently working as senior researcher in the Chemical Processes Department at Abengoa Research, Abengoa.

Contents

Acknowledgment and How to Cite

The editor and publisher thank each of the authors who contributed to this book. The chapters in this book were previously published elsewhere in various formats. To cite the work contained in this book and to view the individual permissions, please refer to the citation at the beginning of each chapter. Each chapter was read individually and carefully selected by the editor; the result is a book that provides a nuanced look at managing the wastewater created through manufacturing in different industries. The chapters included are broken into three sections, which describe the following topics:

- The research group responsible for the study in chapter 1 evaluates and compares the genotoxic and cytotoxic effects of a variety of pesticides, provide a foundational understanding for the issues discussed in chapters 2 and 3.
- In Chapter 2, the editor study the anaerobic decontamination of synthetic wastewater bearing PCP at low temperatures by an EGSB reactor, in order to analyze the inhibition of methanogenesis caused by PCP. This is useful information for the design of future anaerobic wastewater treatment reactors.
- The authors of Chapter 3 offer us insight into the organic phosphorus conversion and removal processes in the presence of periphyton (or similar microbial aggregates), which is important for understanding the biogeochemical circulation of phosphorus in wastewater in order to engineer more effective recovery and reuse technologies.
- In Chapter 4, the authors' research demonstrates that a sequential batch reactor is a flexible and efficient technology for treating slaughterhouse wastewater.
- Based on their research offered here in Chapter 5, Sunny and Mathai propose an integrated design that uses a physicochemical process followed by a biological process as a more efficient treatment for fish processing wastewater, one that would also use less energy than current methods, with reduced sludge as well.
- In Chapter 6, the authors review and summarize current research into high-rate anaerobic treatment for agricultural and food-production wastewater. They note existing problems, and offer some strategies for overcoming them.

- In Chapter 7, the authors investigate the effect of using phenol as cosubstrate, with bioaugmentation with *P. putida*, for biodegradating 4-CP from synthetic wastewater, using sequencing batch reactors. They conclude that because of the high survival rate of this strain and a possible genetic transference with the microorganisms, the sequencing batch reactor is suitable for recalcitrant pollutants.
- In Chapter 8, the authors describe the treatment of wastewater from a specific cosmetic factory located in Spain. They then discuss the drawbacks of the system in use and suggest that membrane batch reactors would be a more cost-efficient treatment technology.
- In chapter 9, Romero and her colleagues investigate the heterogeneous photocatalytic degradation and mineralization of MET in aqueous suspensions with TiO_2, as well as the contributions of direct photolysis and the adsorption of metoprolol onto TiO_2. Their results confirmed that the addition of TiO_2 as a photocatalyst allows a fast and efficient removal of metoprolol.
- The study described in chapter 10 investigates the fate and behavior of antidepressants and their N-desmethyl metabolites following ozone treatment. The authors find that effluent ozonation offers a promising strategy for eliminating antidepressants from wastewater. They acknowledge the potential, however, for this methodology to generate toxic byproducts.
- In chapter 11, the authors propose a methodology for the amidoximation and consequent Fe coordination of dye wastewater with a nanofiber as catalyst. They found that their process showed better catalytic performance than the Fe complex prepared with conventional yarns.
- Bernal and her colleagues describe in chapter 12 their research into the ozonation of dyes catalyzed with Fe-pillared clays. They found that a small amount of the clay substantially enhances the degradation and mineralization of indigo carmine, doubling the noncatalyzed system reaction rate.

List of Contributors

Benoît Barbeau
École Polytechnique de Montréal, Department of Civil, Geological and Mining Engineering, P.O. Box 6079, Succursale Centre-ville, Montreal, Quebec, H3C 3A7, Canada

Carlos Barrera-Díaz
Centro Conjunto de Investigación en Química Sustentable UAEMex-UNAM, Carretera Toluca Atlacomulco Km 14.5, 50200 Toluca, MEX, Mexico

Miriam Bernal
Centro Conjunto de Investigación en Química Sustentable UAEMex-UNAM, Carretera Toluca Atlacomulco Km 14.5, 50200 Toluca, MEX, Mexico

Mireille Blais
École Polytechnique de Montréal, Department of Civil, Geological and Mining Engineering, P.O. Box 6079, Succursale Centre-ville, Montreal, Quebec, H3C 3A7, Canada

Bowen Cheng
School of Textiles, Tianjin Polytechnic University, Tianjin 300387, China and State Key Laboratory of Hollow Fiber Membrane Materials and Processes, Tianjin Polytechnic University, Tianjin 300387, China

Anupam Debsarkar
Environmental Engineering Division, Civil Engineering Department, Jadavpur University, Kolkata 32, India

Yongchun Dong
School of Textiles, Tianjin Polytechnic University, Tianjin 300387, China

Santiago Esplugas
Department of Chemical Engineering, University of Barcelona, C/Martí i Franquès 1, 08028 Barcelona, Spain

Christian Gagnon
Environment Canada, Wastewater and Effluents Section, Water Science and Technology Directorate, 105 McGill Street, Montreal, Quebec, H2Y 2E7, Canada

Jaime Giménez
Department of Chemical Engineering, University of Barcelona, C/Martí i Franquès 1, 08028 Barcelona, Spain

Yung-Tse Hung
Department of Civil and Environmental Engineering, Cleveland State University, Cleveland, OH 44115, USA

Weimin Kang
School of Textiles, Tianjin Polytechnic University, Tianjin 300387, China and State Key Laboratory of Hollow Fiber Membrane Materials and Processes, Tianjin Polytechnic University, Tianjin 300387, China

Pradyut Kundu
Environmental Engineering Division, Civil Engineering Department, Jadavpur University, Kolkata 32, India

André Lajeunesse
Environment Canada, Wastewater and Effluents Section, Water Science and Technology Directorate, 105 McGill Street, Montreal, Quebec, H2Y 2E7, Canada

M. L. Larramendy
Cátedra de Citología, Facultad de Ciencias Naturales y Museo, Universidad Nacional de La Plata, Calle 64 N° 3, B1904AMA La Plata, Argentina and Consejo Nacional de Investigaciones Científicas y Técnicas (CONICET), Argentina

Jesus Lopez
Sección Departamental de Ingeniería Química, Universidad Autónoma de Madrid, C/Francisco Tomás y Valiente 7, 28049 Madrid, Spain

Haiying Lu
State Key Laboratory of Soil and Sustainable Agriculture, Institute of Soil Science, Chinese Academy of Sciences, Nanjing, P. R. China

Pilar Marco
Department of Chemical Engineering, University of Barcelona, C/Martí i Franquès 1, 08028 Barcelona, Spain

Lekha Mathai
Professor, Department of Civil Engineering, M.A. College of Engineering , Kothamangalam, Kerala, India

Angel F. Mohedano
Sección Departamental de Ingeniería Química, Universidad Autónoma de Madrid, C/Francisco Tomás y Valiente 7, 28049 Madrid, Spain

Victor M. Monsalvo
Sección Departamental de Ingeniería Química, Universidad Autónoma de Madrid, C/Francisco Tomás y Valiente 7, 28049 Madrid, Spain

Somnath Mukherjee
Environmental Engineering Division, Civil Engineering Department, Jadavpur University, Kolkata 32, India

Reyna Natividad
Centro Conjunto de Investigación en Química Sustentable UAEMex-UNAM, Carretera Toluca Atlacomulco Km 14.5, 50200 Toluca, MEX, Mexico

N. Nikoloff
Cátedra de Citología, Facultad de Ciencias Naturales y Museo, Universidad Nacional de La Plata, Calle 64 N° 3, B1904AMA La Plata, Argentina and Consejo Nacional de Investigaciones Científicas y Técnicas (CONICET), Argentina

D. Puyol
Sección Departamental de Ingeniería Química, Universidad Autónoma de Madrid, C/Francisco Tomás y Valiente 7, 28049 Madrid, Spain

Rajinikanth Rajagopal
Dairy and Swine Research and Development Centre, Agriculture and Agri-Food Canada, Sherbrooke, Quebec J1M0C8, Canada

Gabriela Roa
Centro Conjunto de Investigación en Química Sustentable UAEMex-UNAM, Carretera Toluca Atlacomulco Km 14.5, 50200 Toluca, MEX, Mexico

Juan J. Rodriguez
Sección Departamental de Ingeniería Química, Universidad Autónoma de Madrid, C/Francisco Tomás y Valiente 7, 28049 Madrid, Spain

Rubí Romero
Centro Conjunto de Investigación en Química Sustentable UAEMex-UNAM, Carretera Toluca Atlacomulco Km 14.5, 50200 Toluca, MEX, Mexico

Violette Romero
Department of Chemical Engineering, University of Barcelona, C/Martí i Franquès 1, 08028 Barcelona, Spain

C. Ruiz de Arcaute
Cátedra de Citología, Facultad de Ciencias Naturales y Museo, Universidad Nacional de La Plata, Calle 64 N° 3, B1904AMA La Plata, Argentina and Consejo Nacional de Investigaciones Científicas y Técnicas (CONICET), Argentina

Noori M. Cata Saady
Dairy and Swine Research and Development Centre, Agriculture and Agri-Food Canada, Sherbrooke, Quebec J1M0C8, Canada

Sébastien Sauvé
Department of Chemistry, Université de Montréal, P.O. Box 6128, Succursale Centre-ville, Montreal, Quebec, H3C 3J7, Canada

S. Soloneski
Cátedra de Citología, Facultad de Ciencias Naturales y Museo, Universidad Nacional de La Plata, Calle 64 N° 3, B1904AMA La Plata, Argentina and Consejo Nacional de Investigaciones Científicas y Técnicas (CONICET), Argentina

Miguel M. Somer
Universidad Autonoma de Madrid, Chemical Engineering Section, Francisco Tomas y Valiente 7, Madrid, 28049, Spain

Neena Sunny
Assistant Professor, Department of Civil Engineering, M.A.College of Engineering , Kothamangalam, Kerala, India

Joseph V. Thanikal
Caledonian Centre for Scientific Research, Department of Built and Natural Environment, Caledonian College of Engineering, P.O. Box 2322, CPO Seeb 111, Muscat, Sultanate of Oman

Montserrat Tobajas
Universidad Autonoma de Madrid, Seccion Departamental de Ingenieria Quimica, Francisco Tomas y Valiente 7, Madrid, 28049, Spain

Teresa Torres-Blancas
Centro Conjunto de Investigación en Química Sustentable UAEMex-UNAM, Carretera Toluca Atlacomulco Km 14.5, 50200 Toluca, MEX, Mexico

Michel Torrijos
INRA, UR50, Laboratory of Environmental Biotechnology, Avenue des Etangs, Narbonne, F-11100, France

Yonghong Wu
State Key Laboratory of Soil and Sustainable Agriculture, Institute of Soil Science, Chinese Academy of Sciences, Nanjing, P. R. China

Linzhang Yang
State Key Laboratory of Soil and Sustainable Agriculture, Institute of Soil Science, Chinese Academy of Sciences, Nanjing, P. R. China and Jiangsu Academy of Agriculture Sciences, Nanjing, P. R. China

Shanqing Zhang
Centre for Clean Environment and Energy, Environmental Futures Centre, Griffith School of Environment, Gold Coast Campus, Griffith University, Queensland, Australia

Xueting Zhao
School of Textiles, Tianjin Polytechnic University, Tianjin 300387, China and State Key Laboratory of Hollow Fiber Membrane Materials and Processes, Tianjin Polytechnic University, Tianjin 300387, China

Introduction

This volume is broken into four sections, each of which examines the treatment of wastewater in a different manufacturing industry. The first section focuses on the phytosanitaries industry; although the benefits generated by the application of agro-chemicals are evident in terms of increased agricultural productivity and improved public health through disease control, the presence of pesticides residues in the environment has created potential risks perspective. Herbicides and insecticides are chemically stable and recalcitrant, and represent a risk to health and the environment, due to their persistent and long-term toxicity. The main causes of water pollution by pesticides includes their use as a routine practice in agriculture and surface runoff, the wash of recipients and dispensing equipments, the manufacture of agricultural products, such as fruits and vegetables washing and disposal of polluted plants. The intensification of agriculture has conducted consequently to an increasing demand of phytosanitaries, which have been found in wastewaters and water reservoirs. To prevent the impact of these compounds it is necessary to develop methods allowing their effective breakdown.

The food and beverage (F&B) industry, the topic of the second section, ranks third in water consumption and wastewater discharge in Europe, after the chemical and refinery industries. Access to water is critical for the F&B industry, both in terms of quantity and quality. The main challenges for the F&B industry in terms of water use are to continuously reduce levels of water consumption in its processes by improving water use and reuse efficiency without compromising strict food hygiene requirements. In the F&B processing sector, water serves two key functions: product and a main ingredient (for bottled water, non-alcoholic and alcoholic drinks, etc.), and its role as heat and mass-transport media in many food-processing steps, such as washing, boiling, steaming, cooling, and cleaning. Fruit, vegetable, dairy, meat, and pre-prepared manufacturers demand tremendous volumes of water for their activities, up to 4.7 m³/ton of product. Besides being essential in processing, water is also the most common waste in the F&B industry, because food processing relies on water for washing, rinsing, boiling, evaporation, extraction, and common waste in the F&B industry, because food processing relies on water for washing, rinsing, boiling, evaporation, extraction, and cleaning.

The third section looks at waste produced by the pharmaceutical industries. Pharmaceutical pollution is now detected in waters throughout the world. Some of this enters our water systems indirectly, from runoff and municipal public septic and water facilities, but pharmaceutical manufacturing plants produce their share of direct water contamination. As a result, cocktail of synthetic compounds, from antidepressants and antibiotics to analgesics and antihistamines, are entering our waterways.

The World Bank estimates that almost 20 percent of global industrial water pollution comes from the treatment and dyeing of textiles, the topic of the final part of this book. Dyeing, rinsing, and treatment of textiles all use large amounts of fresh water—and dyeing makes use of some seventy-two toxic chemicals. Millions of gallons of wastewater discharged by mills each year contain chemicals such as formaldehyde, chlorine, and heavy metals. This wastewater causes chemical and biological changes in our aquatic system, threatening species of fish and aquatic plants, and it makes human water use unhealthy or dangerous. Meanwhile, in many areas of the world, the enormous amount of water required by the textile industry is in direct competition with the growing daily water requirements of the humans who reside in these drought-prone regions.

Victor Monsalvo, PhD

Agrochemicals represent one of the most important sources of environmental pollution. Although attempts to reduce agrochemical use through organic agricultural practices and the use of other technologies to control pests continue, the problem is still unsolved. Recent technological advances in molecular biology and analytical science have allowed the development of rapid, robust, and sensitive diagnostic tests (biomarkers) that can be used to monitor exposure to, and the effects of pollution. One of the major goals Chapter 1, by Larramendy and colleagues, is to evaluate comparatively the genotoxic and cytotoxic effects exerted by several pure agrochemicals and their technical formulations commonly used in Argentina on vertebrate cells in vitro and in vivo employing several end-points for geno and cytotoxicity. Among them are listed the herbicides dicamba and flurochloridone, the fungicide zineb, the insecticides pirimicarb and imidacloprid. Overall, the results clearly demonstrated that the damage induced by the commercial formulations is in general greater than that produced by the pure pesticides, suggesting the presence of deleterious components in the excipients with either a putative intrinsic toxic effect or with the capacity of exacerbating the toxicity of the pure agrochemicals, or both. Accordingly, the results highlight that: 1) A complete knowledge of the toxic effect/s of the active ingredient is not enough in biomonitoring studies; 2) Pesticide/s toxic effect/s should be evaluated assaying to the commercial formulation available in market; 3) The deleterious effect/s of the excipient/s present within the commercial formulation should not be either discarded nor underestimated, and 4) A single bioassay is not enough to characterize the toxicity of an agrochemical under study.

In Chapter 2, Lopez and colleagues investigated the anaerobic treatment of low-strength wastewater bearing pentachlorophenol (PCP) at psychro-mesophilic temperatures in an expanded granular sludge bed reactor. Using an upward flow rate of 4 m h^{-1}, a complete removal of PCP, as well as COD removal and methanization efficiencies higher than 75% and 50%, respectively, were achieved. Methanogenesis and COD consumption were slightly affected by changes in loading rate, temperature (17–28 °C) and inlet concentrations of urea and oils. Pentachlorophenol caused an irreversible inhibitory effect over both acetoclastic and hydrogenotrophic methanogens, being the later more resistant to the toxic effect of pentachlorophenol. An auto-inhibition phenomenon was observed at PCP concentrations higher than 10 mg L^{-1}, which was accurately predicted by a Haldane-like model. The inhibitory effect of PCP over the COD consumption and methane production was modelled by modified pseudo-Monod and Roediger models, respectively.

To understand the role of ubiquitous phototrophic periphyton in aquatic ecosystem on the biogeochemical cycling of organic phosphorus, the conversion and removal kinetic characteristics of organic phosphorus (P_{org})

such as adenosine triphosphate (ATP) were investigated in the presence of the periphyton cultured in artificial non-point source wastewater. The preliminary results in Chapter 3, by Lu and colleagues, showed that the periphyton was very powerful in converting P_{org}, evidenced by the fact that inorganic phosphorus (P_{inorg}) content in solution increased from about 0.7 to 14.3 mg P L^{-1} in 48 hours in the presence of 0.6 g L^{-1} periphyton. This was because the periphyton could produce abundant phosphatases that benefited the conversion of P_{org} to P_{inorg}. Moreover, this conversion process was described more suitable by the pseudo-first-order kinetic model. The periphyton was also effective in removing P_{org}, which showed that the P_{org} can be completely removed even when the initial P_{org} concentration was as high as 13 mg P L^{-1} in 48 hours in the presence of 1.6 g L^{-1} periphyton. Furthermore, it was found that biosorption dominated the P_{org} removal process and exhibited the characteristics of physical adsorption. However, this biosorption process by the periphyton was significantly influenced by biomass (absorbent dosage) and temperature. This work provides insights into P_{org} biogeochemical circulation of aquatic ecosystem that contained the periphyton or similar microbial aggregates.

Slaughterhouse wastewater contains diluted blood, protein, fat, and suspended solids, as a result the organic and nutrient concentration in this wastewater is vary high and the residues are partially solubilized, leading to a very highly contaminating effect in riverbeds and other water bodies if the same is let off untreated. In Chapter 4, Kundu and colleagues investigated the performance of a laboratory-scale Sequencing Batch Reactor (SBR) in aerobic-anoxic sequential mode for simultaneous removal of organic carbon and nitrogen from slaughterhouse wastewater. The reactor was operated under three different variations of aerobic-anoxic sequence, namely, (4+4), (5+3), and (3+5) hr. of total react period with two different sets of influent soluble COD (SCOD) and ammonia nitrogen (NH_4^+-N) level 1000 ± 50 mg/L, and 90 ± 10 mg/L, 1000 ± 50 mg/L and 180 ± 10 mg/L, respectively. It was observed that from 86 to 95% of SCOD removal is accomplished at the end of 8.0 hr of total react period. In case of (4+4) aerobic-anoxic operating cycle, a reasonable degree of nitrification 90.12 and 74.75% corresponding to initial NH_4^+-N value of 96.58 and 176.85 mg/L, respectively, were achieved. The biokinetic coefficients (k,

K_s, Y, k_d) were also determined for performance evaluation of SBR for scaling full-scale reactor in future operation.

The main environmental problems of fish industries are high water consumption and high organic matter, oil and grease, ammonia and salt content in their wastewaters. The generated fish wastewater is rich in oil and grease, salt and ammonia. Biological treatments of such wastewater render them harmless. The primary causes of failure are a wide variety of inhibitory substances present in substantial concentrations in wastes. Chapter 5, by Sunny and Mathai, consequently focuses on the inhibitors of biological treatment process for fish processing wastewater. It could be concluded from studies that system ammonia content, wastewater salinity, oil and grease play a decisive role in the efficiency of fish processing wastewater treatment. Physico-chemical was responsible for biological processes. An integrated design using physicochemical process followed by biological process would yield better treatment efficiency with less energy consumption and reduced sludge production.

Chapter 6, by Rajagopal and colleagues, compiles the various advances made since 2008 in sustainable high-rate anaerobic technologies with emphasis on their performance enhancement when treating agro-food industrial wastewater. The review explores the generation and characteristics of different agro-food industrial wastewaters; the need for and the performance of high rate anaerobic reactors, such as an upflow anaerobic fixed bed reactor, an upflow anaerobic sludge blanket (UASB) reactor, hybrid systems etc.; operational challenges, mass transfer considerations, energy production estimation, toxicity, modeling, technology assessment and recommendations for successful operation.

Chapter 7, by Monsalvo and colleagues, is focused on the application of intensified biological systems for the degradation of halogenated organic compounds. The single and combined effects of cometabolism and bioaugmentation with *Pseudomonas putida* on the aerobic degradation of 4-chlorophenol (4-CP) in sequencing batch reactors was studied. Phenol was added as growth substrate to enhance 4-CP biodegradation through cometabolic transformation. Adaptation of activated sludge by increasing 4-CP loads aimed at a progressive acclimation to that compound, which could be successfully degraded at loading rates below 55 mg g^{-1} VSS d^{-1}. Using phenol as cosusbtrate allowed almost a threefold decrease of the time required for the exhaustion of 4-CP. The addition of phenol also reduced the toxic effect of 4-CP over *P. putida*. The bioaugmentation of the SBR with *P. putida* enhanced the 4-CP removal rate, allowing the SBR to deal with 4-CP loads up to 120 mg g^{-1} VSS d^{-1}. Bioaugmentation of SBR with *P. putida* improves the capacity of this system to withstand high toxic shocks. Cometabolic degradation of 4-CP with phenol improves the removal rates achieved by the SBR at similar 4-CP loads. Both strategies are more convenient intensification techniques than acclimation for the biological treatment of 4-CP.

Cosmetic wastewaters are characterized by relatively high values of chemical oxygen demand (COD), suspended solids, fats, oils and detergents. These effluents have been commonly treated by means of coagulation/flocculation. Nevertheless, the more stringent regulations concerning industrial wastewaters makes necessary to implement new technologies. For this reason, the application of activated carbon adsorption, ultrafiltration and advanced oxidation processes, including catalytic wet peroxide oxidation has been reported in the last years. In Chapter 8, Monsalvo and colleagues describe the treatment of wastewater from a cosmetic factory located in Spain. Initially, the treatment sequence consists of homogenization, filtration, coagulation, neutralization, flocculation, flotation and biological oxidation in a sequencing batch reactor (SBR). The main difficulties for treating cosmetic wastewater by biological processes derive from the presence of detergents, surfactants, hormones, cosmetics and pharmaceutical compounds. Owing to capacity revamping and improvement of the piping, cleaning systems and process units in the last years, the organic loads have increased dramatically with a negative effect on the quality of the effluent. Given the high water needs and the policy adopted in the factory on water recycling as well as the existing limitations in land, a MBR was evaluated as a cost-efficient system.

Chapter 9, by Romero and colleagues, reports the photocatalytic degradation of the ß-blocker metoprolol (MET) using TiO$_2$ suspended as catalyst. A series of photoexperiments were carried out by a UV lamp, emitting in the 250–400 nm range, providing information about the absorption of radiation in the photoreactor wall. The influence of the radiation wavelength on the MET photooxidation rate was investigated using a filter cutting out wavelengths shorter than 280 nm. Effects of photolysis and adsorption at different initial pH were studied to evaluate noncatalytic degradation for this pharmaceutical. MET adsorption onto titania was fitted to two-parameter Langmuir isotherm. From adsorption results it appears that the photocatalytic degradation can occur mainly on the surface of TiO_2. MET removed by photocatalysis was 100% conditions within 300 min, while only 26% was achieved by photolysis at the same time. TiO_2 photocatalysis degradation of MET in the first stage of the reaction followed approximately a pseudo-first-order model. The major reaction intermediates were identified by LC/MS analysis such as 3-(propan-2-ylamino)propane-1,2-diol or 3-aminoprop-1-en-2-ol. Based on the identified intermediates, a photocatalytic degradation pathway was proposed, including the cleavage of side chain and the hydroxylation addition to the parent compounds.

In Chapter 10, Lajeunesse and colleagues examined the fate of 14 antidepressants along with their respective N-desmethyl metabolites and the anticonvulsive drug carbamazepine in a primary sewage treatment plant (STP) and following advanced treatments with ozone (O_3). The concentrations of each pharmaceutical compound were determined in raw sewage, effluent and sewage sludge samples by LC-MS/MS analysis. The occurrence of antidepressant by-products formed in treated effluent after ozonation was also investigated. Current primary treatments using physical and chemical processes removed little of the compounds (mean removal efficiency: 19%). Experimental sorption coefficients (K_d) of each studied compounds were also calculated. Sorption of venlafaxine, desmethylvenlafaxine, and carbamazepine on sludge was assumed to be negligible (log $K_d \leq 2$), but higher sorption behavior can be expected for sertraline (log $K_d \geq 4$). Ozonation treatment with O_3 (5 mg/L) led to a satisfactory mean removal efficiency of 88% of the compounds. Screening of the final ozone-treated effluent samples by high resolution-mass spectrometry (LC-QqToFMS) did confirm the presence of related N-oxide by-products. Effluent ozonation led to higher mean removal efficiencies than current primary treatment, and therefore represented a promising strategy for the elimination of antidepressants in urban wastewaters. However, the use of O_3 produced by-products with unknown toxicity.

In Chapter 11, Zhao and colleagues prepared the modified PAN nanofiber Fe complex by the amidoximation and Fe coordination of PAN nanofiber was obtained using electrospinning technique and then used for the heterogeneous Fenton degradation of textile dyes as a novel catalyst. Some main factors affecting dye degradation such as Fe content of catalyst, irradiation intensity, H_2O_2 initial concentration, the solution pH as well as dye structure, and initial concentration were investigated. UV-Vis spectrum analysis and TOC measurement were also used to evaluate the dye degradation process. The results indicated that the modified PAN nanofiber Fe complex exhibited a much better catalytic activity for the heterogeneous Fenton degradation of textile dyes than the Fe complex prepared with conventional PAN yarns in the dark or under light irradiation. Increasing Fe content of catalyst or irradiation intensity would accelerate the dye degradation. And the highest degradation efficiency was obtained with $3.0 \, mmol \, L^{-1} \, H_2O_2$ at pH 6. Moreover, this complex was proved to be a universal and efficient catalyst for degradation of three classes of textile dyes including azo dye, anthraquinone dye, and triphenylmethane dye. Additionally, the dye mineralization was also significantly enhanced in the presence of this complex.

Chapter 12, by Bernal and colleagues, studied the ozonation catalyzed by iron-pillared clays. The degradation of dye indigo carmine (IC) was elected as test reaction. Fe-pillared clays were synthesized by employing hydrolyzed $FeCl_3$ solutions and bentonite. The pillared structure was verified by XRD and by XPS the oxidation state of iron in the synthesized material was established to be +2. By atomic absorption the weight percentage of iron was determined to be 16. The reaction was conducted in a laboratory scale up-flow bubble column reactor. From the studied variables the best results were obtained with a particle size of 60 microns, pH = 3, ozone flow of 0.045 L/min, and catalyst concentration of 100 mg/L. IC was completely degraded and degradation rate was found to be double when using Fe-PILCS than with ozone alone. DQO reduction was also significantly higher with catalyzed than with noncatalyzed ozonation.

PART I

GROWING OUR FOOD:
THE PHYTOSANITARIES INDUSTRY

CHAPTER 1

Genotoxicity and Cytotoxicity Exerted by Pesticides in Different Biotic Matrices: An Overview of More Than a Decade of Experimental Evaluation

M. L. LARRAMENDY, N. NIKOLOFF, C. RUIZ DE ARCAUTE, AND S. SOLONESKI

1.1 PROBLEM FRAMEWORK

Nowadays, it is worldwide accepted that the survival of humans as a species is intimately linked to the well-being of ecosystems and the resources they can provide. However, it is also well assume that the well-being of ecosystems depends, in turn, on minimizing the damaging impacts of anthropogenic activities. Irrespective of the kinds of habitats we choose to protect or restore, we need to understand how ecosystems, and the organisms that inhabit them, respond to chemicals exposure, among other detrimental factors. Recent technological advances in molecular biology

Genotoxicity and Cytotoxicity Exerted by Pesticides in Different Biotic Matrices-An Overview of More Than a Decade of Experimental Evaluation. © *Larramendy ML, Nikoloff N, Ruiz de Arcaute C, and Soloneski S.* Journal of Environmental & Analytica Toxicology **4,**225 (2014), doi: 10.4172/2161-0525.1000225. *Licensed under Creative Commons Attribution License, http://creativecommons.org/licenses/by/3.0.*

and analytical science have allowed the development of rapid, robust, and sensitive diagnostic tests (biomarkers) to monitor both exposure and the effects of pollutants. For the first time, we are able to make health assessments of individual organisms in much the same way that we evaluate human health.

It is estimated that approximately 1.8 billion people worldwide engage in agriculture and most use pesticides to protect the food and commercial products that they produce. Others use pesticides occupationally for public health programs, and in commercial applications, while many others use pesticides for lawn and garden applications and in and around the home [1,2]. Pesticides are defined as "chemical substance or mixture of substances used to prevent, destroy, repel or mitigate any pest ranging from insects (i.e., insecticides), rodents (i.e., rodenticides), and weeds (i.e., herbicides) to microorganisms (i.e., algicides, fungicides, and bactericides)" [1,3,4]. Definition of pesticide varied with times and countries. Nevertheless, the essence of pesticide has remained and remains basically constant, i.e., it is a (mixed) substance that is poisonous and efficient to target organisms and is safe to non-target organisms and environments.

Years ago, it has been reported that more than 2,000,000 million tn of pesticides are used only in the US each year whereas approximately over 11,000,000 million tn are used worldwide [1]. However, it is very well known that in many developing countries programs to control exposures are limited or even non-existent. Therefore, it has been estimated that among living species worldwide, only as many as 25 million agricultural workers experience unintentional pesticide poisonings each year [5]. According to the WHO [6] unintentional poisonings kill an estimated 355,000 people globally each year. In developing countries, where two thirds of these deaths occur, such poisonings are associated strongly with excessive exposure to, and inappropriate use of, toxic chemicals. Furthermore, the OECD has estimated that by the year 2020, nearly one third of the world's chemical production will take place in non-OECD countries and that global output will be 85% higher than it was in 1995. Therefore, the chemical shift of production from developed countries to poor countries could cause an increase in both the risks of environmental health in the second category of countries [7].

Although attempts to reduce pesticide use through organic agricultural practices and the use of other technologies to control pests continue, exposure to pesticides occupationally, through home and garden use, through termite control or indirectly through spray drifts and through residues in household dust, and in food and water are common [8-14]. The US Department of Agriculture has estimated that 50 million people in the US obtain their drinking water from groundwater that is potentially contaminated by pesticides and other agricultural chemicals [9,15-26]. Children from 3-6 years old received most of their dermal and non-dietary oral doses from playing with toys and while playing on carpets which contributed the largest portion of their exposure [22-27].

In epidemiological and in experimental biology studies, the existence of an increasing interest in biomonitoring markers to achieve both a measurement and an estimation of biologically active/passive exposure to genotoxic pollutants, is nowadays a real fact. Significant contributions to the advancement of pesticide toxicology came and continue to come from many sources, e.g., academic, governmental/ regulatory, and industrial. Regulatory agencies, private sector, and academia worldwide combine expertise to assess pesticide safety and risk potential demanding adequate data of high quality to serve as the basis for establishing safe exposure levels. The extent of testing was and is often determined by the depth of the science, as well as the chemical and physical properties of the agent and the extent of exposure. The importance of pesticide toxicology has evolved from listing poisons to protecting the public from the adverse effects of chemicals, from simply identifying effects (qualitative toxicology), to identifying and quantifying human risks from exposure (quantitative toxicology), and from observing phenomena to experimenting and determining mechanisms of action of pesticide agents and rational management for intoxication. Humans and living species may, therefore, be exposed to a number of different chemicals through dietary and other routes of exposure.

Pesticides are ubiquitous on the planet and they are employed to control or eliminate a variety of agricultural and household pests that can damage crops and livestock and to enhance the productivity. Despite the many benefits of the use of pesticides in crops field and its significant contribution to the lifestyles we have come to expect, pesticides can also be hazardous if not used appropriately and many of them may represent potential hazards due to the contamination of food, water, and air, which can result in severe health problems not only for humans but also for ecosystems [28]. The actual number of pesticide-related illnesses is unknown, since many poisonings go unreported. It has been estimated that at least three million cases of pesticide poisoning occur worldwide each year (www.who.int). The majority of these poisonings occur in developing countries where less protection against exposure is achieved, knowledge of health risks and safe use is limited or even unknown. Studies in developed countries have demonstrated the annual incidence intoxication in agricultural workers can reach values up to 182 per million and 7.4 per million among full time workers [29] and schoolchildren [30], respectively. However, the number of poisonings increases dramatically in emerging countries where the marketing of pesticides is often uncontrolled or illicit and the misbranded or unlabelled formulations are sold at open stands (www.who.int). Yet, cases of pesticide intoxication may be the result of various causes in different regions of the world. In emerging countries, where there is insufficient regulation, lack of surveillance systems, less enforcement, lack of training, inadequate or reduced access to information systems, poorly maintained or nonexistent personal protective equipment's, and larger agriculturally based populations, the incidences are expected, then, to be higher [31]. Despite the magnitude of the problem of pesticide poisoning, there have been very few detailed studies around the world to identify the risk factors involved with their use. The use of pesticides banned in industrialized countries, in particular, highly toxic pesticides as classified by WHO, US EPA, and IARC, obsolete stockpiles and improper storage techniques may provide unique risks in the developing world, where 25% of the global pesticide production is consumed [28]. Particularly, the impact of increased deregulation of agrochemicals in Latin America threatens to increase the incidence of pesticide poisoning, which has already been termed a serious public health problem throughout the continent by the WHO. Many of the pesticides used in Latin America are US exports and the companies can make a number of changes to ensure the "safe" use of their products. However, the social, economic and cultural conditions under which they are used, pesticides acutely poison hundreds of thousands each year, including many children.

There is an aspect related with use and misuse of pesticides that should be commented on further. The continuous subtoxic exposures of these agrochemicals raises the concern about which is the behavior, environmental fate and the potential adverse effects on both target and non target organisms once incorporated into the environment. The different chemical products used in agriculture could be distributed within the environment by means of drift, surface runoff, and drainage [32,33] and, thus, can be found far away from the point of application. The mobility of pesticides in soil and hence their transfer to other environmental compartments, depends on a variety of complex dynamic physical, chemical and biological processes, including sorption–desorption, volatilization, chemical and/or biological degradation, uptake, runoff, and leaching, among other factors [34-37]. In addition, many pesticides can persist for long periods in the ecosystem. Furthermore, once a persistent pesticide has entered the food chain, it can undergo "biomagnification", i.e., accumulation in the body tissues of organisms, where it may reach concentrations many times higher than in the surrounding environment and directly compromising the health of organisms, including humans [38-40].

In the majority of Latin American countries, poisoning registries are so inadequate that most acute poisoning cases never get recorded. Meanwhile, health effects of chronic or long-term pesticide exposures such as cancer or birth defects are not available, omissions that serve to hide the epidemic proportion of pesticide-related illness in the region. In Argentina, e.g., available official data revealed that 79% of the intoxications due to pesticides are related with the use of herbicides followed by insecticides and fungicides (www.msal.gov.ar), values that correlate with the evolution of the phytosanitary market demonstrating that herbicides accounted for the largest portion of total use (69%), followed by insecticides (13%), and fungicides (11%) (www.casafe.org). Consequently, Argentina a larger producer of cereals, including soy, is actually the world eight-largest agrochemical market. The country has seen an explosion in genetically modified soybean production with soy exports topping $16.5 billion in 2008 (www.casafe.org). The fertile South American nation is now the world's third largest producer of soy, trailing behind the United States and Brazil.

Furthermore, there is an aspect that should be further considered. It is well known that in agriculture, pesticides are usually applied in their formulated forms, where the active ingredient is combined with organic solvents and emulsifying and wetting agents, which affect the pesticide penetration and performance [41]. The additives may synergize or antagonize the toxicity of the active ingredient. However, additive compounds frequently make up part of a commercial pesticide formulation, they are not usually included in any discussion of the effects on living organisms, and their adverse effects may exceed those of the active ingredient. Although pesticides are developed through very strict regulation processes to function with reasonable certainty and minimal impact on human health and the environment, serious concerns have been raised about health risks resulting from occupational exposure and from residues in food and drinking water [41]. Several investigations have demonstrated that the additive compounds present in pesticide commercial formulations have the ability to induce cellular toxicity, including genotoxicity and genotoxicity by themselves, separate from the active ingredient [42-51]. Accordingly, risk assessment must also consider additional toxic effects caused by the excipient(s). Thus, both the workers as well as non-target organisms are exposed to the simultaneous action of the active ingredient and a variety of other chemical/s contained in the formulated product.

Since more than a decade, one of the major goals of our research group has been to evaluate comparatively the genotoxic and cytotoxic effects exerted by several pure pesticides Pestanal® analytical standards (Riedel-de Haën, Germany) and their technical formulations commonly used in Argentina on eukaryotic cells employing several biotic matrices both in vitro and in vivo. Among them are included the herbicides dicamba and the 57.7% dicamba-based formulation Banvel® (Syngenta Agro S.A., Buenos Aires, Argentina) and flurochloridone and the 25.0% flurochloridone-based formulations Twin Pack Gold® (Magan Argentina, S.A., Buenos Aires, Argentina) and Rainbow® (Syngenta Agro S.A., Buenos Aires, Argentina), the fungicide zineb and the 70.0% zineb-based formulation Azzurro® (Chemiplant, Buenos Aires, Argentina), and the insecticides pirimicarb and the 50.0% pirimicarb-based formulations Aficida® (Syngenta Agro S.A., Buenos Aires, Argentina) and Patton Flow® (Gleba S.A., Buenos Aires, Argentina). For the particular case of the insecticide imi-

dacloprid, the 35.0% imidacloprid-based formulation Glacoxan imida®
(Punch Química S.A., Buenos Aires, Argentina) was assayed. The sister
chromatid exchange (SCE), cell-cycle progression (CCP), structural chromosome aberrations (CA), single cell gel electrophoresis assay (SCGE),
spindle disturbances, micronuclei (MN), mitotic index (MI), MTT, and
neutral red (NR) bioassays were used as end-points for geno and cytotoxicity in several cell systems including in vitro nontransformed and transformed mammalian cells, and in vivo *Allium cepa* meristematic root cells
as well as circulating blood cells from *Rhinella arenarum* (Anura, Bufonidae) and *Hypsiboas pulchellus* (Anura, Hylidae) tadpoles. The aforementioned agrochemicals were chosen because they represent one of the most
employed pesticides used for pest control not only in Argentina but also
worldwide scale. A simple search within the Farm Chemical International
database clearly reveals this concept (www.farmchemicalsinternational.
com). So far, whereas available information indicates the existence of 34
basic producers and eight formulators for dicamba, six basic producers
and at least two formulators worldwide are related with the manufacture
and marketing of the herbicide flurochloridone. For the fungicide zineb,
it has been reported the existence of 21 and at least seven basic producers
and formulators, respectively. Finally, at global scale, the existence of 19
basic producers and at least four formulators as well as 117 basic producers and at least 49 formulators are related with the manufacture and marketing of the insecticides pirimicarb and imidacloprid.

1.2 DICAMBA: GENOTOXICITY AND CYTOTOXICITY PROFILES

Dicamba (3,6-dichloro-2-methoxybenzoic acid; CASRN: 1918-00- 9)
is a selective systemic herbicide, absorbed by the leaves and roots, acts
as an auxin-like growth regulator causing uncontrolled growth [52]. It is
used to control annual and perennial broad-leaved weeds and bush species, e.g. cereals, maize, sorghum, sugar cane, asparagus, perennial seed
grasses, turf, pastures, rangeland, and non-crop land [52]. Based on its
acute toxicity, dicamba has been classified as a class II member (moderately hazardous) by WHO (http://www.who.int/ipcs/publications/pes-
ticides hazard/en/) and slightly to moderately toxic (category II-III) by

US EPA [52]. Genotoxicity and cytotoxicity investigations have been conducted with this auxinic member using several end-points on different cellular systems. When mutagenic activity was assessed in bacterial systems with the *Salmonella typhimurium* Ames test either positive or negative results have been reported [53-55]. Furthermore, similar situation were observed in *Escherichia coli* and *Bacillus subtilis* when the reverse mutation assay was applied [53,56,57]. Whereas the herbicide was unable to induce mitotic recombination on *Saccharomyces cerevisiae* [58], negative and positive results were obtained for the induction of unscheduled DNA synthesis in human primary fibroblasts regardless of the presence or absence of S9 mix [53,59]. Sorensen et al. [60,61] found positive results on dicamba-treated CHO-K1 cells cultured in the presence of reducedclay smectites but not when the clay system were not included within the culture protocol. Perocco et al. [59] demonstrated the ability of the herbicide to induce SCEs in CHO-K1 cells and human lymphocytes in vitro with and without S9 fraction, respectively. It has been reported the ability of the herbicide to give positive results by using the gene mutation and recombination assays when *Arabidopsis thaliana* was used as experimental model [62]. However, both negative and inconclusive results were reported for the sex-linked recessive lethal mutation end-point in dicamba-exposed *Drosophila melanogaster* [57,63]. Perocco and co-workers [59] reported an increased frequency of DNA unwinding rate in rat hepatocytes. It has been also reported that the herbicide is able to enhance the frequency of CA in the rootand hoot-tip cells of barley and in rat bone marrow cells [64]. Finally, Mohamed and Ma [65] reported the MN induction in *Tradescantia sp.*

In our laboratory, we have studied the genotoxicity and cytotoxicity in vitro of the herbicide dicamba and the dicambacontaining commercial formulation Banvel® in human lymphocytes as well as in CHO-K1 cells (Figure 1). We were able to demonstrate that dicamba is a DNA-damaging agent since enhancement of the frequency of SCEs (Figure 1A), MN (Figure 1C), and single DNA strand breaks (Figure 1B) in mammalian in vitro cells [66,67]. Similarly, we demonstrated the induction of alterations in the CCP (Figure 1E), reduction of the MI status (Figure 1D), and cell viability after in vitro dicamba and Banvel® exposure [66-68].

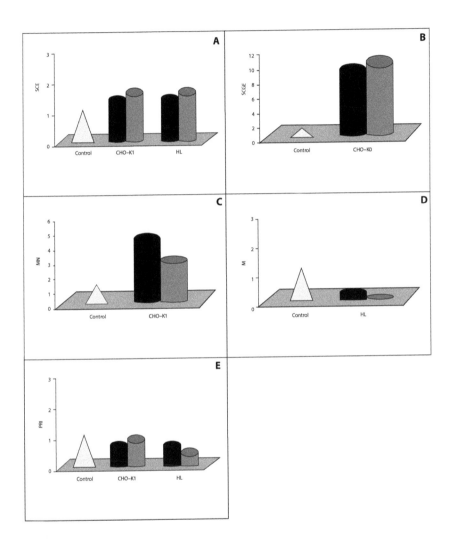

FIGURE 1: Comparative genotoxicity and cytotoxicity effects induced by dicamba (black cylinders) and the dicamba-based herbicide formulation Banvel® (grey cylinders) commonly used in Argentina on in vitro mammalian Chinese hamster ovary (CHO-K1) cells and human lymphocytes (HL). Results are expressed as fold-time values over control data (white pyramid). Evaluation was performed using end-points for genotoxicity [SCEs (A), SCGE (B), and MN (C)] and cytotoxicity [MI (D) and PRI (E)].

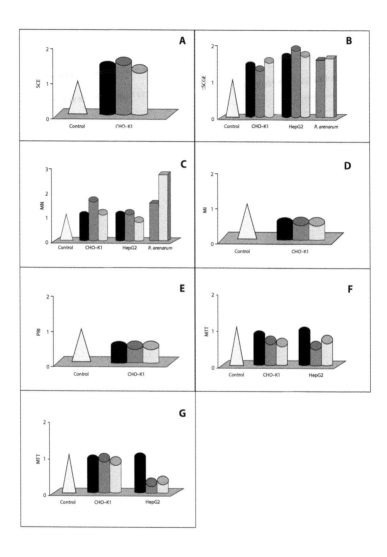

FIGURE 2: Comparative genotoxicity and cytotoxicity effects induced by flurochloridone (black) and the flurochloridone-based herbicide formulations Twin Pack Gold® (dark grey) and Rainbow® (light grey) commonly used in Argentina on in vitro (cylinders) mammalian Chinese hamster ovary (CHO-K1) and human hepatocellular carcinoma (HepG2) cells and in vivo (prisms) circulating blood *R. arenarum* tadpole cells. Results are expressed as foldtime values over control data (white pyramid). Evaluation was performed using end-points for genotoxicity [SCEs (A), SCGE (B), and MN (C)] and cytotoxicity [MI (D), PRI (E), MTT (F), and NR (G)].

1.3 FLUROCHLORIDONE: GENOTOXICITY AND CYTOTOXICITY PROFILES

Flurochloridone (3-chloro-4-(chloromethyl)-1-[3-(trifluoromethyl)phen-yl]-2-pyrrolidinone; CASRN: 89286-81-7) is a pre-emergence herbicide used to control a range of weeds in umbelliferous, cereal, sunflower, and potato crops, among others [69]. Toxicological information for flurochloridone has been poorly documented. So far, it has been reported that the herbicide does not reveal genotoxic, carcinogenic, or neurotoxic potential in rodents [69]. The herbicide induces low or moderate acute toxicity in rats when administered by oral, dermal, or inhalational routes [69]. However, it causes adverse effects in male reproductive functions and hormonal system alterations [69]. Accessible information on the genotoxic properties of flurochloridone is scarce. To the best of our knowledge, a single report has been reported so far. When root meristematic cells of *Allium cepa* were exposed to the herbicide, abnormal CCP and cellular mitodepressive activity were found [70]. The most frequently observed abnormalities were c-metaphases, multipolarity, polyploidy, and chromosome lagging. In addition, chromosomal stickiness, chromosome breaks, bridges, fragments, sister union, and MN were also observed after flurochloridone exposure [70].

Recently, we demonstrated that both flurochloridone and its formulations Twin Pack Gold® and Rainbow® are DNA-damaging agents (Figure 2), since an enhancement of the frequency of SCEs (Figure 2A), alterations in lysosomal (Figure 2G) and mitochondrial activities (Figure 2F), a delay in the CCP (Figure 2E) as well as a decrease of the MI (Figure 2D) were observed to occur in in vitro treated mammalian CHO-K1 cells [48]. Furthermore, by using the same in vitro cellular system, we recently demonstrated the ability of flurochloridone to induce DNA single-strand breaks (Figure 2B) and MN frequency (Figure 2C) [47]. Similarly, both flurochloridone and the flurochloridone-based formulation were able to exert the same genotoxic and cytotoxic pattern on HepG2 cells in vitro (Figures 2B,C), hepatocellular carcinoma cell line maintaining phase I and II enzymes [71]. Finally, when the MN induction (Figure 2C) and DNA strand breaks (Figure 2B) estimation by the SCGE assay were employed as in vivo end-points, positive results were reported in erythrocytes of Twin Pack Gold®- and Rainbow®-exposed *R. arenarum* tadpoles by Nikoloff and collaborators [72].

1.4 ZINEB: GENOTOXICITY AND CYTOTOXICITY PROFILES

Zineb (ethylene bis(dithiocarbamate) zinc; CASRN: 12122- 67-7) is a widely employed foliar fungicide with prime agricultural and industrial applications [73]. Although zineb has been mainly registered to be used on a large number of fruits, vegetables, field crops, ornamental plants, and for the treatment of seeds, it has also been registered to be used as a fungicide in paints and for mold control on fabrics, leather, linen, painted surfaces, surfaces to be painted, and on paper, plastic, and wood surfaces [73]. It has been classified as a compound practically nontoxic (class IV) by US EPA [73] based on its potency by the oral and inhalation exposure routes. The available data on the deleterious effects of zineb do not allow a definitive evaluation of its carcinogenic potential and it has been not classified as to its carcinogenicity to humans (category III) by IARC [74]. This fungicide alters thyroid hormone levels and/or weights. The reproductive system is generally unaffected after zineb exposure [73].

Genotoxicity and cytotoxicity studies have been conducted with this dithiocarbamate member using several end-points on different cellular matrices. Zineb have been generally recognized as non-mutagenic in bacteria, yeast and fungi as well as in mammalian cells [73]. Plate incorporation assay with *S. typhimurium* demonstrated a direct non-mutagenic effect of the fungicide whereas mitotic chromosome malsegregation, gene conversion and point mutation assays with *S. cerevisiae* and *B. subtilis* gave positive results [75, 76]. Tripathy et al. [77] reported zineb as positive genotoxic agent to somatic and germ cells in *Drosophila* sp. While Chernov and Khitsenko [78] observed an increased incidence of lung tumors after its oral administration to C57BL mice, negative results have been also reported to occur either in other mouse strains [79] or in rats [80]. A variety of sarcomas were observed after subcutaneous administration in mice and rats [81]. Also, Enninga and coworkers [82] showed that zineb induced structural CA in CHO cells both with and without S9. In contrast to these studies, it was reported that the fungicide did not induce MN in bone marrow cells of Wistar male rats after oral administration [83]. In humans, haemolytic alterations have been reported after zineb contact [84]. Finally, an increase in the frequency of CA was observed in the lymphocytes of persons occupationally exposed to zineb [85]. Several assays have been developed to assess the ability of zineb to cause cytotoxic effects on different cellular systems. Zineb exerted a high dose-related cytotoxicity in BALB/c 3T3 mouse cells in vitro but only in the absence of an exogenous metabolizing system [86]. However, Whalen and coworkers [87] reported negative results when human natural killer cells were exposed to zineb. However, alterations in the mitochondrial transmembrane potential and cardiolipin content were reported to occur after zineb administration in rats [88].

We evaluated comparatively the genotoxic and cytotoxic in vitro effects induced in vitro by the pure fungicide and its commercial formulation Azzurro® on CHO-K1 cells, human non-transformed fibroblast and circulating lymphocytes as well as on in vivo *A. cepa* meristematic root cells (Figure 3). Our observations revealed the ability of both zineb and the zineb-based formulation to induce CA in human lymphocytes (Figure 3D) [89,90]. Similarly, the fungicide increased the frequency of SCEs (Figure 3A) and modified the CCP (Figure 3F) and the MI status (Figure 3E) on human lymphocytes and CHO-K1 cells [89,90]. We have also demonstrated that both zineb and Azzurro® were not only able to induce MN in human lymphocytes in vitro, but also that such induction was restricted to B CD20+ and T suppressor/ cytotoxic CD8+ cell subsets [91]. Furthermore, when assessing DNA damage and repair kinetics analyzed using the SCGE assay on zineband Azzurro®-CHO-K1 exposed cells, we observed that single strand breaks introduced into the DNA molecule likely reflect those induced by alkylating agents rather than those produced by active oxygen species (Figure 3B) [92]. Finally, we have also observed using a β-tubulin immunodetection assay that the exposure to Azzurro® interferes with normal assembly of microtubule structures during the mitosis of A. cepa meristematic root cells [93] and in mammalian transformed and non-transformed exposed cell lines [94].

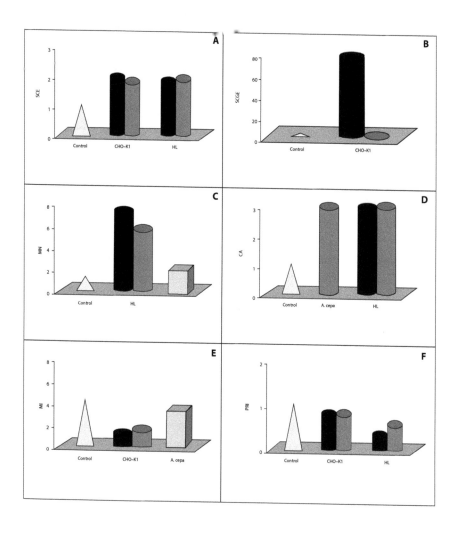

FIGURE 3: Comparative genotoxicity and cytotoxicity effects induced by zineb (black) and the zineb-based fungicide formulation Azzurro® (dark grey) commonly used in Argentina on in vitro (cylinders) mammalian Chinese hamster ovary (CHO-K1) cells and human lymphocytes (HL) and in vivo (prisms) *A. cepa* meristematic root cells. Results are expressed as fold-time values over control data (white pyramid). Evaluation was performed using end-points for genotoxicity [SCEs (A), SCGE (B), MN (C), and CA (D)] and cytotoxicity [MI (E) and PRI (F)].

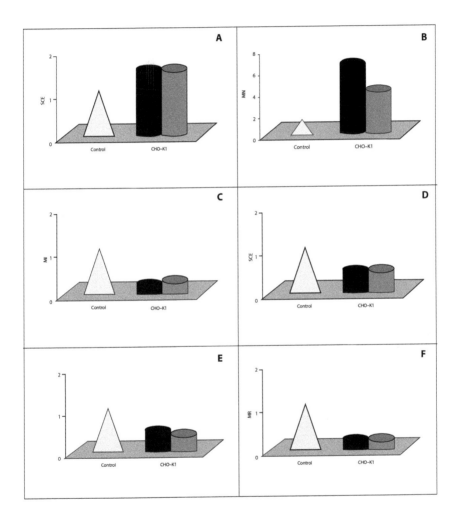

FIGURE 4: Comparative genotoxicity and cytotoxicity effects induced by pirimicarb (black) and the pirimicarb-based insecticide formulation Aficida® (dark grey) commonly used in Argentina on in vitro (cylinders) mammalian Chinese hamster ovary (CHO-K1) cells. Results are expressed as fold-time values over control data (white pyramid). Evaluation was performed using end-points for genotoxicity [SCEs (A) and CA (B)] and cytotoxicity [MI (C), PRI (D), MTT (E), and NR (F)].

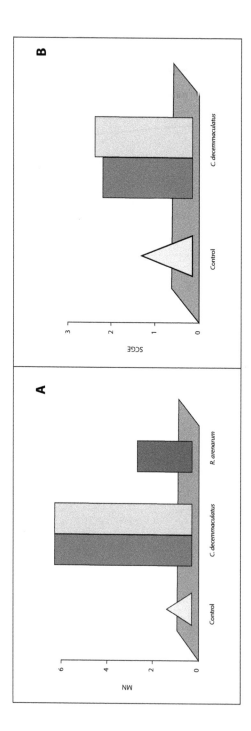

FIGURE 5: Comparative genotoxicity and cytotoxicity effects induced by pirimicarb (black) and the pirimicarb-based insecticide formulations Aficida® (dark grey) and Patton Flow® (light grey) commonly used in Argentina on in vivo (prism) circulating blood *R. arenarum* tadpole and *C. decemmaculatus* cells. Results are expressed as fold-time values over control data (white pyramid). Evaluation was performed using end-points for genotoxicity [MN (A) and SCGE (B)].

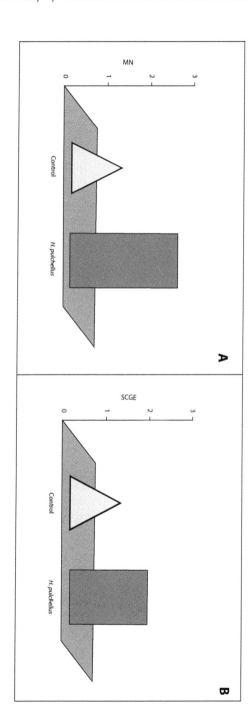

FIGURE 6: Comparative genotoxicity and cytotoxicity effects induced by imidacloprid (black) and the imidacloprid-based insecticide formulation Glacoxan Imida® (dark grey) commonly used in Argentina on in vivo (prism) circulating blood *H. pulchellus* tadpole cells. Results are expressed as fold-time values over control data (white pyramid). Evaluation was performed using end-points for genotoxicity [MN (A) and SCGE (B)].

1.5 PIRIMICARB: GENOTOXICITY AND CYTOTOXICITY PROFILES

Pirimicarb(2-dimethylamino-5,6-dimethylpyrimidin-4- yldimethylcarba-mate; CASRNı 23103-98-2) is a derivative of carbamic acid insecticide member with both contact and systemic activity. Based on its acute toxicity, pirimicarb has been classified as a moderately hazardous compound (class II) by WHO [95] and slightly to moderately toxic (category II-III) by US EPA [96]. Pirimicarb is registered as a fast-acting selective aphicide mostly used in a broad range of crops, including cereals, sugar beet, potatoes, fruit, and vegetables, and is relatively non-toxic to beneficial predators, parasites, and bees [28,97]. Its mode of action is inhibiting acetylcholinterase activity [28,97].

Available information on the genotoxic and cytotoxic properties of pirimicarb is limited and inconsistent. Only few data are available in the literature [28,97]. Genotoxicity and cytotoxicity studies have been conducted with this carbamate using several end-points on different cellular systems. Pirimicarb has been generally recognized as nongenotoxic in bacteria, yeast and fungi as well as in mammalian cells [28,97]. It has been reported to be non-mutagenic in bacteria systems [98,99]. Negative and positive results were obtained for the induction of mutagenicity in mouse lymphoma L5178Y cells regardless of the presence or absence of S9 mix [100]. Furthermore, evaluation of the induction of DNA single strand breaks revealed positive results in human lymphocytes exposed in vitro [101]. It has been reported the ability of the insecticide to give positive results by using the eye mosaic system *white/white+* (w/w+) somatic mutation and recombination test (SMART) when *D. melanogaster* was employed [102]. However, others authors reported negative results when mutation bioassays was performed in rats [103,104]. At the chromosomal level, pirimicarb did not induce CA in bone marrow cells of rats after oral administration [105,106]. Contrarily, Pilinskaia [107] observed a significant increase of CA in the peripheral blood lymphocytes from occupational workers after pirimicarb exposure.

We evaluated comparatively the genotoxic and cytotoxic in vitro effects induced by the pure insecticide and its commercial formulation Aficida® on in vitro CHO-K1 cells (Figure 4) as well as on in vivo biotic matrices including the fish *C. decemmaculatus* and amphibian *R. arena-*

rum tadpoles (Figure 5). Our observations revealed positive results for both compounds results when the either the CA (Figure 4B) and the SCE (Figure 4A) assays were performed in CHO-K1 cells [51]. Furthermore, the induction of alterations in the CCP (Figure 4D) and MI status (Figure 4C) on CHO-K1 cells was reported to occur after in vitro exposure to pirimicarb [51]. Finally, when the MN induction (Figure 5A), alterations in the erythrocytes:erythroblasts ratios, and SCGE end-points (Figure 5B) were employed after in vivo exposure to the pirimicarb-based formulations Aficida® and Patton Flow®, positive results were reported by Vera Candioti and collaborators in *C. decemmaculatus* [108,109] and *R. arenarum* tadpoles exposed under laboratory conditions [110].

1.6 IMIDACLOPRID: GENOTOXICITY AND CYTOTOXICITY PROFILES

Imidacloprid, (2E)-1-[(6-Chloro-3-pyridinyl)methyl]-N-nitro- 2-imidazolidinimine; CASRN: 138261-41-3), is a nicotine-derived systemic insecticide belonging to the neonicotinoids pesticide group. These insecticides act as an insect neurotoxin and belongs to a class of chemicals, chloronicotinyl nitroguanidine chemical family, which affect the central nervous system of insects [111,112]. It is effective on contact and via stomach action (http://extoxnet.orst.edu/pips/imidaclo.htm). Because imidacloprid binds much more strongly to insect nicotinic neuron receptors than that of mammal neurons, this insecticide results selectively more toxic to insects than mammals [112,113]. Imidacloprid has been ranked as a class II chemical (moderately hazardous) by the WHO [114] whereas the US EPA [115] has included the insecticide into the Group E of compounds with no evidence of carcinogenicity.

Imidacloprid decreases the reproduction rates in *Caenorhabditis elegans* and *Eisenia fetida* [116]. After S9 metabolic activation in vitro, imidacloprid produces calf thymus DNA adducts [117], increases the frequency of spermatic abnormalities in *E. fetida* [118], and is mutagenic in *S. typhimurium* strains, with or without S9 fraction [119]. The insecticide also induces significant increases in the frequency of SCE and MN formation in human peripheral blood lymphocytes [120,121], mice and rat bone-marrow cells [119,122], peripheral blood erythrocytes from *Rana N-Hallowell* tadpoles [123], and *Vicia faba* root cells [118]. Furthermore, imidacloprid causes DNA strand breaks in the coelomocytes of *E. fetida* [118], erythrocytes from *Rana N-Hallowell* anuran tadpoles [123], human peripheral blood lymphocytes [120], and leukocytes in vitro [121]. However, it does not cause DNA strand breaks in *V. faba* root cells [123].

In our laboratory, we have recently studied the in vivo genotoxic effects induced by the imidacloprid-based commercial formulation Glacoxan imida® on *H. pulchellus* tadpoles exposed under laboratory conditions (Figure 6). Our observations demonstrated that the insecticide is able to exert DNA and chromosomal damage evaluated by the MN (Figure 6A) and SCGE (Figure 6B) bioassays [124].

1.7 FINAL REMARKS

Overall, a comparative analysis of results revealed, depending upon the end-point employed, that the damage induced by the commercial formulations of the pesticides is, in general and regardless of the type of the active ingredient, greater than that produced by the pure compounds by themselves. Unfortunately, the identity of the components present within the excipient formulations was not made available by the manufacturer. These final remarks are in accord with previous observations not only reported by us but also by other research groups indicating the presence of xenobiotics within the composition of the commercial formulations with genotoxic and cytotoxic effects as previously mentioned [44,46,51,66-68,89,90,125-130]. Hence, risk assessment must also consider additional genocytotoxic effects caused by the excipient/s. Thus, both the workers as well as non-target organisms are exposed to the simultaneous action of the active ingredient and a variety of other chemical/s contained in the formulated product.

Finally, the results highlight that a whole knowledge of the toxic effect/s of the active ingredient of a pesticide is not enough in biomonitoring studies as well as that agrochemical/s toxic effect/s should be evaluated according to the commercial formulation available in market. Furthermore, the deleterious effect/s of the excipient/s present within the commercial formulation should be neither discarded nor underestimated. The importance of further studies on this type of pesticide in order to achieve a complete knowledge on its genetic toxicology seems to be, then, more than evident.

REFERENCES

1. Donaldson D, Kiely T, Grube AU (2002) Pesticide's industry sales and usage 1998-1999 market estimates, Environmental Protection Agency; Washington (DC): Report No. EPA-733-R-02-OOI.
2. Repetto R, Baliga S (1996) Trends and patterns of pesticide use. Public Health Risks, 3-8, in: Pesticides and the immune system, World Resources Institute, Washington, D.C.
3. Alavanja MC (2009) Introduction: pestic ides use and exposure extensive worldwide. Rev Environ Health 24: 303-309.
4. USEPA (2009) What is a Pesticide?US Environmental Protection Agency, Washington, DC.
5. Jeyaratnam J (1990) Acute pesticide poisoning: a major global health problem. World Health Stat Q 43: 139-144.
6. WHO (2003) Shaping the future of global health, Bulletin of The World Health Report, World Health Organization: 81.
7. OECD (2001) Environmental outlook for the chemicals industry. Environment Directorate, Organization for Economic Co-operation and Development, World Health Organization.
8. Zolgharnein J, Shahmoradi A, Ghasemi J (2011) Pesticides removal using conventional and low-cost adsorbents: a review. Clean-Soil, Air, Water39: 1105-1119.
9. Damalas CA, Eleftherohorinos IG (2011) Pesticide exposure, safety issues, and risk assessment indicators. Int J Environ Res Public Health 8: 1402-1419.
10. Tankiewicz M, Fenik J, Biziuk M (2010) Determination of organophosphorus and organonitrogen pesticides in water samples. Trends Anal Chem29: 1050-1063.
11. Baris RD, Cohen SZ, Barnes NL, Lam J, Ma Q (2010) Quantitative analysis of over 20 years of golf course monitoring studies. Environ ToxicolChem 29: 1224-1236.
12. Hernández F, Sancho JV, Ibáñez M, Grimalt S (2008) Investigation of pesticide metabolites in food and water by LC-TOF-MS. Trends Anal Chem27: 862-872.
13. Schipper PN, Vissers MJ, van der Linden AM (2008) Pesticides in groundwater and drinking water wells: overview of the situation in the Netherlands. Water SciTechnol 57: 1277-1286.
14. Hamilton DJ, Ambrus Ã, Dieterle RM, Felsot AS, Harris CA, et al. (2003) Regulatory limits for pesticide residues in water (IUPAC technical report). Pure ApplChem75: 1123-1155.
15. Lindahl AML, Bockstaller C (2012) An indicator of pesticide leaching risk to groundwater. Ecol Indic23: 95-108.

16. Cabeza Y, Candela L, Ronen D, Teijon G (2012) Monitoring the occurrence of emerging contaminants in treated wastewater and groundwater between 2008 and 2010. The BaixLlobregat (Barcelona, Spain). J Hazard Mater 239-240: 32-9.

17. Morgenstern U, Daughney CJ (2012) Groundwater age for identification of baseline groundwater quality and impacts of land-use intensification - The National Ground-water Monitoring Programme of New Zealand. J Hydrol456-457: 79-93.

18. González S, López-Roldán R, Cortina JL (2012) Presence and biological effects of emerging contaminants in Llobregat River basin: a review. Environ Pollut 161: 83-92.

19. Stuart M, Lapworth D, Crane E, Hart A (2012) Review of risk from potential emerging contaminants in UK groundwater. Sci Total Environ 416: 1-21.

20. Kumar M, Puri A (2012) A review of permissible limits of drinking water. Indian J Occup Environ Med 16: 40-44.

21. Diduch M, Polkowska Z, NamieÂ›nik J (2011) Chemical quality of bottled waters: a review. J Food Sci 76: R178-196.

22. Vonderheide AP, Bernard CE, Hieber TE, Kauffman PE, Morgan JN, et al. (2009) Surface-to-food pesticide transfer as a function of moisture and fat content. J Expo Sci Environ Epidemiol 19: 97-106.

23. Jurewicz J, Hanke W, Johansson C, Lundqvist C, Ceccatelli S, et al. (2006) Adverse health effects of children's exposure to pesticides: what do we really know and what can be done about it. ActaPaediatrSuppl 95: 71-80.

24. Rohrer CA, Hieber TE, Melnyk LJ, Berry MR (2003) Transfer efficiencies of pesticides from household flooring surfaces to foods. J Expo Anal Environ Epidemiol 13: 454-464.

25. Lewis RG, Fortune CR, Blanchard FT, Camann DE (2001) Movement and deposition of two organophosphorus pesticides within a residence after interior and exterior applications. J Air Waste ManagAssoc 51: 339-351.

26. Akland GG, Pellizzari ED, Hu Y, Roberds M, Rohrer CA, et al. (2000) Factors influencing total dietary exposures of young children. J Expo Anal Environ Epidemiol 10: 710-722.

27. Lu C, Fenske RA (1999) Dermal transfer of chlorpyrifos residues from residential surfaces: Comparison of hand press, hand drag, wipe, and polyurethane foam roller measurements after broadcast and aerosol pesticide applications. Environ Health Perspect107: 463-467.

28. WHO-FAO (2009) Pesticides residues in food, in: FAO Plant Production and Protection Paper. Rome 1-426.

29. Calvert GM, Plate DK, Das R, Rosales R, Shafey O, et al. (2004) Acute occupational pesticide-related illness in the US, 1998-1999: surveillance findings from the SENSOR-pesticides program. Am J Ind Med 45: 14-23.

30. Alarcon WA, Calvert GM, Blondell JM, Mehler LN, Sievert J, et al. (2005) Acute illnesses associated with pesticide exposure at schools. JAMA 294: 455-465.

31. IFCS (2003) Acutely toxic pesticides: initial input on extent of problem and guidance for risk management. Fourth session of the Intergovernmental Forum on Chemical Safety. Doc number: IFCS/FORUM-IV/10, Bangkok.

32. Meffe R, de Bustamante I2 (2014) Emerging organic contaminants in surface water and groundwater: a first overview of the situation in Italy.Sci Total Environ 481: 280-295.

33. Aparicio VC, De Gerónimo E, Marino D, Primost J, Carriquiriborde P, et al. (2013) Environmental fate of glyphosate and aminomethylphosphonic acid in surface waters and soil of agricultural basins. Chemosphere 93: 1866-1873.

34. Vryzas Z, Papadopoulou-Mourkidou E, Soulios G, Prodromou K (2007) Kinetics and adsorption of metolachlor and atrazine and the conversion products (deethylatrazine, deisopropylatrazine, hydroxyatrazine) in the soil profile of a river basin. Eur J Soil Sci: 58 1186-1199.

35. Roy A, Singh SK, Bajpai J, Bajpai AK (2014) Controlled pesticide release from biodegradable polymers. CEJC12: 453-469.

36. Li WC (2014) Occurrence, sources, and fate of pharmaceuticals in aquatic environment and soil. Environ Pollut 187: 193-201.

37. Zaki MS, Hammam AM (2014) Aquatic pollutants and bioremediations. Life Sci J11: 362-369.

38. Zhang K, Wei YL, Zeng EY (2013) A review of environmental and human exposure to persistent organic pollutants in the Pearl River Delta, South China. Sci Total Environ 463-464: 1093-110.

39. Manciocco A, Calamandrei G2, Alleva E3 (2014) Global warming and environmental contaminants in aquatic organisms: the need of the etho-toxicology approach. Chemosphere 100: 1-7.

40. Liu G, Cai Z, Zheng M (2014) Sources of unintentionally produced polychlorinated naphthalenes. Chemosphere 94: 1-12.

41. WHO (1990) Public health impacts of pesticides used in agriculture (WHO in collaboration with the United Nations Environment Programme, Geneva, 1990), World Health Organization.

42. Belden J, McMurry S, Smith L, Reilley P (2010) Acute toxicity of fungicide formulations to amphibians at environmentally relevant concentrations. Environ ToxicolChem 29: 2477-2480.

43. Brühl CA, Schmidt T, Pieper S, Alscher A (2013) Terrestrial pesticide exposure of amphibians: an underestimated cause of global decline? Sci Rep 3: 1135.

44. Lin N, Garry VF (2000) In vitro studies of cellular and molecular developmental toxicity of adjuvants, herbicides, and fungicides commonly used in Red River Valley, Minnesota. J Toxicol Environ Health A 60: 423-439.

45. Mann RM, Bidwell JR (1999) The toxicity of glyphosate and several glyphosate formulations to four species of southwestern Australian frogs. Arch Environ ContamToxicol 36: 193-199.

46. Molinari G, Kujawski M, Scuto A, Soloneski S, Larramendy ML (2012) DNA damage kinetics and apoptosis in ivermectin-treated chinese hamster ovary cells. J ApplToxicol .

47. Nikoloff N, Larramendy ML, Soloneski S (2012) Comparative evaluation in vitro of the herbicide flurochloridone by cytokinesis-block micronucleus cytome and comet assays. Environ Toxicol .

48. Nikoloff N, Soloneski S, Larramendy ML (2012) Genotoxic and cytotoxic evaluation of the herbicide flurochloridone on Chinese hamster ovary (CHO-K1) cells. ToxicolIn Vitro 26: 157-163.

49. Rayburn AL, Moody DD, Freeman JL (2005) Cytotoxicity of technical grade versus formulations of atrazine and acetochlor using mammalian cells. Bull Environ ContamToxicol 75: 691-698.

50. Zeljezic D, Garaj-Vrhovac V, Perkovic P (2006) Evaluation of DNA damage induced by atrazine and atrazine-based herbicide in human lymphocytes in vitro using a comet and DNA diffusion assay. ToxicolIn Vitro 20: 923-935.

51. Soloneski S, Larramendy ML (2010) Sister chromatid exchanges and chromosomal aberrations in Chinese hamster ovary (CHO-K1) cells treated with the insecticide pirimicarb. J Hazard Mater 174: 410-415.

52. USEPA (1974) Compendium of Registered Pesticides. US Government Printing Office, Washington, DC.

53. Simmon VF (1979) In vitro microbiological mutagenicity and unscheduled DNA synthesis studies of eighteen pesticides. In: EPA-600/1-79-04, EPA, Research Triangle Park 1-79.

54. Plewa MJ, Wagner ED, Gentile GJ, Gentile JM (1984) An evaluation of the genotoxic properties of herbicides following plant and animal activation. Mutat Res136: 233-245.

55. Kier LD, Brusick DJ, Auletta AE, Von Halle ES, Brown MM, et al. (1986) The Salmonella typhimurium/mammalian microsomal assay: A report of the U.S. Environmental Protection Agency Gene-Tox Program. Mutat Res 168: 69-240.

56. Leifer Z, Kada T, Mandel M, Zeiger E, Stafford R, et al. (1981) An evaluation of tests using DNA repair-deficient bacteria for predicting genotoxicity and carcinogenicity. A report of the U.S. EPA's Gene-TOX Program.Mutat Res 87: 211-297.

57. Waters MD, Nesnow S, Simmon VF, Mitchell AD, Jorgenson TA, et al. (1981) Pesticide Chemist and Modern Toxicology. American Chemical Society, Washington, DC.

58. Zimmermann FK, von Borstel RC, von Halle ES, Parry JM, Siebert D, et al. (1984) Testing of chemicals for genetic activity with Saccharomyces cerevisiae: a report of the U.S. Environmental Protection Agency Gene-Tox Program. Mutat Res 133: 199-244.

59. Perocco P, Ancora G, Rani P, Valenti AM, Mazzullo M, et al. (1990) Evaluation of genotoxic effects of the herbicide dicamba using in vivo and in vitro test systems. Environ Mol Mutagen 15: 131-135.

60. Sorensen KC, Stucki JW, Warner RE, Plewa MJ (2004) Alteration of mammalian-cell toxicity of pesticides by structural iron(II) in ferruginous smectite. Environ Sci-Technol 38: 4383-4389.

61. Sorensen KC, Stucki JW, Warner RE, Wagner ED, Plewa MJ (2005) Modulation of the genotoxicity of pesticides reacted with redox-modified smectite clay. Environ Mol Mutagen 46: 174-181.

62. Filkowski J, Besplug J, Burke P, Kovalchuk I, Kovalchuk O (2003) Genotoxicity of 2,4-D and dicamba revealed by transgenic Arabidopsis thaliana plants harboring recombination and point mutation markers. Mutat Res 542: 23-32.

63. Lee WR, Abrahamson S, Valencia R, von Halle ES, Wurgler FE, et al. (1983) The sex-linked recessive lethal test for mutagenesis in Drosophila melanogaster. A report of the U.S. Environmental Protection Agency Gene-Tox Program, Mutat Res 123: 183-279.

64. Hrelia P, Vigagni F, Maffei F, Morotti M, Colacci A, et al. (1994) Genetic safety evaluation of pesticides in different short-term tests. Mutat Res 321: 219-228.

65. Mohammed KB, Ma TH (1999) Tradescantia-micronucleus and -stamen hair mutation assays on genotoxicity of the gaseous and liquid forms of pesticides. Mutat Res 426: 193-199.

66. González NV, Soloneski S, Larramendy ML (2006) Genotoxicity analysis of the phenoxy herbicide dicamba in mammalian cells in vitro. ToxicolIn Vitro 20: 1481-1487.

67. González NV, Soloneski S, Larramendy ML (2007) Thechlorophenoxy herbicide dicamba and its commercial formulation banvel induce genotoxicity and cytotoxicity in Chinese hamster ovary (CHO) cells. Mutat Res 634: 60-68.

68. González NV, Soloneski S, Larramendy ML (2009) Dicamba-induced genotoxicity in Chinese hamster ovary (CHO) cells is prevented by vitamin E. J Hazard Mater 163: 337-343.

69. EFSA (2010) Peer review report to the conclusion regarding the peer review of the pesticide risk assessment of the active substance flurochloridone. EFSA Journal8: 1869-1935.

70. Yüzbasioglu D, Ünal F, Sancak C, Kasap R (2003) Cytological effects of the herbicide racer "flurochloridone" on Allium cepa. Caryologia56: 97-105.

71. Nikoloff N, Larramendy ML, Soloneski S2 (2014) Assessment of DNA damage, cytotoxicity, and apoptosis in human hepatoma (HepG2) cells after flurochloridone herbicide exposure. Food ChemToxicol 65: 233-241.

72. Nikoloff N, Natale GS, Marino D, Soloneski S, Larramendy ML (2014) Flurochloridone-based herbicides induced genotoxicity effects on Rhinellaarenarum tadpoles (Anura: Bufonidae). Ecotoxicol Environ Saf100: 275-281.

73. USEPA (1996) Pesticide Fact Sheet: Zineb. US Government Printing Office, Washington, DC.

74. IARC (1976) Some Carbamates, Thiocarbamates and Carbazides. International Agency for Research on Cancer, Lyon.

75. Della Croce C, Morichetti E, Intorre L, Soldani G, Bertini S, et al. (1996) Biochemical and genetic interactions of two commercial pesticides with the monooxygenase system and chlorophyllin. J Environ PatholToxicolOncol 15: 21-28.

76. Franekic J, Bratulic N, PavlicaM, Papes D (1994) Genotoxicity of dithiocarbamates and their metabolites. Mutat Res 325: 65-74.

77. Tripathy NK, Dey L, Majhi B, Das CC (1988) Genotoxicity of zineb detected through the somatic and germ-line mosaic assays and sex-linked recessive-lethal test in Drosophila melanogaster. Mutat Res206: 25-31.

78. Chernov OV, Khitsenko II (1969) Blastomogenic properties of some derivatives of dithiocarbamic acid, VopOnkol: 15 71-74.

79. Innes JR, Ulland BM, Valerio MG, Petrucelli L, Fishbein L, et al. (1969) Bioassay of pesticides and industrial chemicals for tumorigenicity in mice: a preliminary note. J Natl Cancer Inst 42: 1101-1114.

80. SMITH RB Jr, FINNEGAN JK, LARSON PS, SAHYOUN PF, DREYFUSS ML, et al. (1953) Toxicologic studies on zinc and disodium ethylene bisdithiocarbamates. J PharmacolExpTher 109: 159-166.

81. NTIS NTIS(1968) Evaluation of Carcinogenic, Teratogenic and Mutagenic Activities of Selected Pesticides and Industrial Chemicals.United States Department of Commerce, Washington, DC.

82. Enninga IC (1986) Evaluation of the mutagenic activity of zineb in the Ames Salmonella/microsome test. Unpublished report dated. December 29, 1986 from Pennwalt Holland B.V.

83. Centre HR (1985) Reverse mutation/ Salmonella typhimurium. Report No. FMT4/85394. Unpublished report dated August 23, 1985 from Huntingdon Research Centre, England.

84. Pinkhas J, Djaldetti M, Joshua H, Resnick C, de Vries A (1963) Sulphemoglobinemia and acute hemolytic anemia with Heinz bodies following contact with a fungicide - zinc ethylene bisdithiocarbamate- in a subject with glucose-6-phosphatase dehydrogenase deficiency and hypocatalasemia. Blood21: 484-494.

85. Pilinskaia MA (1974) Results of cytogenetic examination of persons occupationally contacting with the fungicide zineb. Genetika10: 140-146.

86. Perocco P, Colacci A, Bonora B, Grilli S (1995) In vitro transforming effect of the fungicides metalaxyl and zineb. TeratogCarcinog Mutagen 15: 73-80.

87. Whalen MM, Loganathan BG, Yamashita N, Saito T (2003) Immunomodulation of human natural killer cell cytotoxic function by triazine and carbamate pesticides. ChemBiol Interact 145: 311-319.

88. Astiz M, de Alaniz MJ, Marra CA (2009) Effect of pesticides on cell survival in liver and brain rat tissues.Ecotoxicol Environ Saf 72: 2025-2032.

89. Soloneski S, González M, Piaggio E, Apezteguía M, Reigosa MA, et al. (2001) Effect of dithiocarbamate pesticide zineb and its commercial formulation azzurro. I. Genotoxic evaluation on cultured human lymphocytes exposed in vitro. Mutagenesis16: 487-493.

90. Soloneski S, González M, Piaggio E, Reigosa MA, Larramendy ML (2002) Effect of dithiocarbamate pesticide zineb and its commercial formulation azzurro. III. Genotoxic evaluation on Chinese hamster ovary (CHO) cells. Mutat Res514: 201-212.

91. Soloneski S, Reigosa MA, Larramendy ML (2002) Effect of dithiocarbamate pesticide zineb and its commercial formulation, azzurro. II. micronucleus induction in immunophenotyped human lymphocytes. Environ Mol Mutagen 40: 57-62.

92. González M, Soloneski S, Reigosa MA, Larramendy ML (2003) Effect of dithiocarbamate pesticide zineb and its commercial formulation, azzurro. IV. DNA damage and repair kinetics assessed by single cell gel electrophoresis (SCGE) assay on Chinese hamster ovary (CHO) cells. Mutat Res 534: 145-154.

93. Andrioli NB, Soloneski S, Larramendy ML, Mudry MD (2012) Cytogenetic and microtubule array effects of the zineb-containing commercial fungicide formulation Azzurro(®) on meristematic root cells of Allium cepa L. Mutat Res 742: 48-53.

94. Soloneski S, Reigosa MA, Larramendy ML (2003) Effect of dithiocarbamate pesticide zineb and its commercial formulation azzurro. V. Abnormalities induced in the spindle apparatus of transformed and non-transformed mammalian cell lines. Mutat Res536: 121-129.

95. WHO (1988) The WHO recommended classification of pesticides by hazard and guidelines to the classification 1988-1989. World Health Organization, Geneva 1-200.

96. USEPA (1974) Pesticide Fact Sheet: Pirimicarb, in, US Government Printing Office, Washington, DC.

97. WHO-FAO (2004) Pesticides residues in food-2004, in: FAO Plant Production and Protection paper World Health Organization and Food and Agriculture Organization of the United Nations, Rome 154-161.

98. Callander RD (1995) Pirimicarb: an evaluation of the mutagenic potential using S. typhimurium and E. coli. Central Toxicology Laboratory Report No. CTL/P/4798 GLP, Unpublished.

99. Käfer E, Scott BR, Dorn GL, Stafford R (1982) Aspergillusnidulans: systems and results of tests for chemical induction of mitotic segregation and mutation. I. Diploid and duplication assay systems. A report of the U.S. EPA Gene-Tox Program, Mutat Res: 98 1-48.

100. Clay P (1996) Pirimicarb: L5178Y TK+/- Mouse Lymphoma Mutation Assay. Central Toxicology Laboratory. Report No: CTL/P/5080 GLP, Unpublished.

101. UndeÄŸer U, BaÀŸaran N (2005) Effects of pesticides on human peripheral lymphocytes in vitro: induction of DNA damage. Arch Toxicol 79: 169-176.

102. Aguirrezabalaga I, Santamaría I, Comendador MA (1994) The w/w+ SMART is a useful tool for the evaluation of pesticides. Mutagenesis 9: 341-346.

103. McGregor DB (1974) Dominant lethal study in mice of ICI PP062. Zeneca unpublished report No. CTL/C/256 from Inveresk Research International. Submitted to WHO by Syngenta Crop Protection AG. Conducte according to OECD 478 (1983). GLP compliant.

104. Kennelly JC (1990) Pirimicarb: Assessment for the induction of unscheduled DNA synthesis in rat hepatocytes in vivo. Unpublished report No. CTL/P/2824 from Central Toxicology Laboratory, Zeneca. Submitted to WHO by Syngenta Crop Protection AG. Conducte according to OECD 486 (1983).GLP compliant.

105. Anderson D, Richardson CR, Howard CA, Bradbrook C, Salt MJ (1980) Pirimicarb: a cytogenetic study in the rat. World Health Organization.

106. Jones K, Howard CA (1989) Pirimicarb (technical): an evaluation in the mouse micronucleus test. Unpublished report No. CTL/P/2641 from Central Toxicology Laboratory, Zeneca. Submitted to WHO by Syngenta Crop Protection AG. Conducted according to OECD 474 (1983).GLP compliant.

107. Pilinskaia MA (1982) [Cytogenetic effect of the pesticide pirimor in a lymphocyte culture of human peripheral blood in vivo and in vitro]. Tsitol Genet 16: 38-42.

108. Candioti JV, Soloneski S, Larramendy ML (2010) Genotoxic and cytotoxic effects of the formulated insecticide Aficida on Cnesterodondecemmaculatus (Jenyns, 1842) (Pisces: Poeciliidae). Mutat Res 703: 180-186.

109. Vera-Candioti J, Soloneski S, Larramendy ML (2013) Pirimicarb-based formulation-induced genotoxicity and cytotoxicity on the fresh water fish Cnesterodondecemmaculatus (Jenyns, 1842) (Pisces, Poeciliidae). ToxicolInd Health.

110. Vera-Candioti J, Natale GS, Soloneski S, Ronco AE, Larramendy ML (2010) Sublethal and lethal effects on Rhinellaarenarum (Anura, Bufonidae) tadpoles exerted by the pirimicarb-containing technical formulation insecticide Aficida®. Chemosphere 78: 249-255.

111. Blacquière T, Smagghe G, van Gestel CA, Mommaerts V (2012) Neonicotinoids in bees: a review on concentrations, side-effects and risk assessment. Ecotoxicology 21: 973-992.

112. Tomizawa M, Casida JE (2005) Neonicotinoid insecticide toxicology: mechanisms of selective action. Annu Rev PharmacolToxicol 45: 247-268.

113. Gervais JA, Luukinen B, Buhl K, Stone D (2010) Imidacloprid Technical Fact Sheet. National Pesticide Information Center.Oregon State University Extension Services.

114. WHO (2002) The WHO recommended classification of pesticides by hazard and guidelines to the classification 2000-2002. World Health Organization, Geneva1-58.

115. NPIC (2010) Imidacloprid. Technical Fact Sheet.National Pesticide Information Center, Oregon State University Extension Services.

116. Gomez-Eyles JL, Svendsen C, Listci L, Martin H, Hodson ME, et al. (2009) Measuring and modelling mixture toxicity of imidacloprid and thiacloprid on Caenorhabditiselegans and Eiseniafetida. Ecotoxicol Environ Saf 72: 71-79.

117. Shah RG, Lagueux J, Kapur S, Levallois P, Ayotte P, et al. (1997) Determination of genotoxicity of the metabolites of the pesticides Guthion, Sencor, Lorox, Reglone, Daconil and Admire by 32P-postlabeling. Mol Cell Biochem 169: 177-184.

118. Zang Y, Zhong Y, Luo Y, Kong ZM (2000) Genotoxicity of two novel pesticides for the earthworm, Eiseniafetida. Environ Pollut 108: 271-278.

119. Karabay NU, Oguz MG (2005) Cytogenetic and genotoxic effects of the insecticides, imidacloprid and methamidophos. Genet Mol Res 4: 653-662.

120. Feng S, Kong Z, Wang X, Peng P, Zeng EY (2005) Assessing the genotoxicity of imidacloprid and RH-5849 in human peripheral blood lymphocytes in vitro with comet assay and cytogenetic tests. Ecotoxicol Environ Saf 61: 239-246.

121. Costa C, Silvari V, Melchini A, Catania S, Heffron JJ, et al. (2009) Genotoxicity of imidacloprid in relation to metabolic activation and composition of the commercial product. Mutat Res 672: 40-44.

122. Demsia G, Vlastos D, Goumenou M, Matthopoulos DP (2007) Assessment of the genotoxicity of imidacloprid and metalaxyl in cultured human lymphocytes and rat bone-marrow. Mutat Res 634: 32-39.

123. Feng S, Kong Z, Wang X, Zhao L, Peng P (2004) Acute toxicity and genotoxicity of two novel pesticides on amphibian, Rana N. Hallowell. Chemosphere 56: 457-463.

124. Pérez-Iglesias JM, Ruiz de Arcaute C, Nikoloff N, Dury L, Soloneski S, et al. (2014) The genotoxic effects of the imidacloprid-based insecticide formulation GlacoxanImida on Montevideo tree frog Hypsiboaspulchellus tadpoles (Anura, Hylidae). Ecotoxicol Environ Saf104: 120-126.

125. Soloneski S, Reigosa MA, Molinari G, González NV, Larramendy ML (2008) Genotoxic and cytotoxic effects of carbofuran and furadan on Chinese hamster ovary (CHOK1) cells. Mutat Res 656: 68-73.

126. Soloneski S, Reigosa MA, Larramendy ML (2003) Effect of the dithiocarbamate pesticide zineb and its commercial formulation, the azzurro. V. Abnormalities induced in the spindle apparatus of transformed and non-transformed mammalian cell lines. Mutat Res 536: 121-129.

127. Molinari G, Soloneski S, Reigosa MA, Larramendy ML (2009) In vitro genotoxic and cytotoxic effects of ivermectin and its formulation ivomec on Chinese hamster ovary (CHOK1) cells. J Hazard Mater 165: 1074-1082.

128. Elsik CM, Stridde HM, Tann RS (2008) Glyphosate adjuvant formulation with glycerin. In: ASTM Special Technical Publication 53-58.

129. David D (1982) Influence of technical and commercial decamethrin, a new synthetic pyrethroid, on the gonadic germ population in Quail embryos. Arch AnatHistolEmbryol 65: 99-110.

130. Cox C, Surgan M (2006) Unidentified inert ingredients in pesticides: implications for human and environmental health. Environ Health Perspect 114: 1803-1806.

CHAPTER 2

Low-Temperature Anaerobic Treatment of Low-Strength Pentachlorophenol-Bearing Wastewater

J. LOPEZ, V. M. MONSALVO, D. PUYOL, A. F. MOHEDANO, AND J. J. RODRIGUEZ

2.1 INTRODUCTION

Pentachlorophenol (PCP) is a biocide used as wood preservative. It was extensively used as a wide spectrum pesticide until its prohibition in 2004 by the Rotterdam Convention. Large quantities of PCP are also produced from ECF-type paper pulp bleaching, the most commonly used nowadays (Savant et al., 2006). Due to its toxic character, PCP is of environmental concern since it acts on a variety of organisms as a potent inhibitor of the oxidative phosphorylation. It disrupts the proton gradient across membranes, interfering with energy transduction of cells (Chen et al., 2008). Moreover, its carcinogenic potential has been recently demonstrated (Cooper and Jones, 2008). The use of this chemical over decades, its resistance to biodegradation, its potential bioaccumulation and biomagnification (Letcher et al., 2009) have led to a wide spread in the environment. PCP has been identified as a priority hazardous substance by the European Community. The environmental quality standards for surface water established by the Directive 2000/60/EC limit the average and maximum allowable concentrations to 0.4 and a 1.0 µg L^{-1}, respectively.

Some tropical countries have continued using PCP as pesticide and component of several agrochemicals in agricultural areas, which is the main cause of its presence in surface and underground water, where concentrations up to 20 mg L^{-1} have been detected in some cases (Downs et al., 1999). Although its contribution as environmental pollutant has declined, PCP still appears in water reservoirs even in countries where this chemical has been banned for several years (Damianovic et al., 2009), reaching concentrations up to 0.36 µg L^{-1} (Gasperi et al., 2008). The presence of PCP in urban areas is thought to be caused through its leakage to the municipal sewer system as a result of accidental spills and dumps. Recent studies have identified the contribution of PCP to contaminant loads into domestic wastewater, reaching concentrations between 0.02 and 0.05 µg L^{-1} in domestic greywater (Nielsen et al., 2005). Treatment by conventional systems appears to be insufficient, since pesticides have been detected in resulting effluents. Thus, highly-efficient removal from wastewater is needed to prevent potential human health risks associated with the ingestion over long periods.

Chlorophenols have been commonly removed from effluents by adsorption, which transfers the problem to the spent adsorbent which becomes a hazardous waste. On the opposite, chemical and biological destructive methods allow the mineralization of chlorophenols. Although chemical oxidation methods are faster than biological ones, the potential generation of byproducts eventually more toxic than the starting pollutants can be a serious problem. The toxic effect of PCP in biological reactors leads to persistent-inherent and persistent-readily biodegradability for aerobic and anaerobic biological systems, respectively. Thus, the combination of both chemical and biological processes has been proposed to achieve high removal efficiencies (Essam et al., 2007).

TABLE 1: Experiences on anaerobic biological treatment of PCP in lab scale reactors.

Reactor[a]	PCPLR (mg PCP L⁻¹ d⁻¹)	PCP (mg L⁻¹)	HRT (d)	OLR (g COD L⁻¹ d⁻¹)	T (°C)	Co-substrate[b]	Chlorophenols detected	Refs.
UASB	97	60	NS	16.2	28	Met, Ace, Pro	NS	Wu et al. (1993)
UASB	0.3	1.0	2.0–3.0	0.4	35	Ph	DCP	Duff et al. (1995)
UASB-b	32	NS	0.2–1.0		NS	Lac	345TCP	Christiansen and Ahring (1996)
GAC-FBBR	0.4	1333	9.3	6.3	35	Eth	3CP, 4CP	Khočadoust et al. (1997)
FBBR	23	1.9	0.1	0.2	35	Glu or But	4CP, 2CP	Mohn et al. (1999)
FBBR	218	100	NS	NS	35	Eth	4CP, 3CP, 34DCP, 35DCP, Ph	Koran et al. (2001)
UASB	62	NS	1.2	0.9	35	Suc, But, Eth	345TCP, 35DCP, 3CP	Tartakovsky et al. (2001)
Hybrid reactor	16.8	21	1.3	6.9	32–37	Met, Ace, Pro, But	NS	Montenegro et al. (2001)
ASBR	5	5000	NS	5.0	28	Suc	NS	Ye and Shen (2004)
FFBR	313	364	1.2–1.5	4.5–5.6	35	Suc, But, Eth, YE	3CP	Lanthier et al. (2005)
UASB	181	152.0	1.0	12.0	28	Suc	NS	Shen et al., 2005 and Shen et al., 2006
HAIB	2–13	2–13	1.0	1.15	30	Glu, Ace, For, Eth	NS	Faraldi et al. (2008)
ASBR	17.7	4.4	2.0	NS	24–32	Suc	345TCP	Mun et al. (2008)
HAIB	4.1	8.0	0.7	1.7	30	Glu, Ace, For, Eth	DCP	Damianovic et al. (2009)

NS: Not specified. [a]UASB-b: UASB reactor with biofilms; GAC-FBBR: FBBR with granular activated carbon; HAIB: Horizontal-flow anaerobic immobilized biomass reactor; ASBR: Anaerobic sequencing batch reactor. [b]Ace: acetate; But: butyrate; Eth: ethanol; For: formiate; Glu: glucose; Lac: lactate; Met: methanol; Pep: peptone; Ph: phenol; Pro: propionate; Suc: sucrose; YE: yeast extract.

Several works have reported on the anaerobic treatment of wastewaters containing chlorinated compounds by different technologies (Farhadian et al., 2008), including anaerobic bioreactors with immobilized biomass (Khodadoust et al., 1997, Lanthler et al., 2005 and Montenegro et al., 2001), fluidized bed biofilm reactors (FBBR) (Mun et al., 2008), sequencing batch reactors (SBR) (Mun et al., 2008), up-flow anaerobic sludge blanket reactors (UASB) (Duff et al., 1995, Shen et al., 2005 and Wu et al., 1993), and membrane anaerobic bioreactors have been recently presented as an effective technology to treat toxic compounds (Singhania et al., 2012). Table 1 shows the experimental conditions used for the treatment of PCP in lab scale anaerobic reactors and the byproducts detected in the resulting effluents. Most of the works dealing with anaerobic PCP removal have shown that low PCP concentrations (0.75–11.3 μM) inhibit methanogenesis completely (Duff et al., 1995 and Wu et al., 1993). Juteau et al. (1995) reported 99% PCP removal in a fixed-film biological reactor (FFBR) with a PCP load of 60 μM d^{-1}. The major intermediates were 3,4-dichlorophenol (3,4-DCP) and 3-chlorophenol (3-CP) and no significant 3-CP degradation was observed. Damianovic et al. (2009) treated efficiently PCP loads up to 15.6 μM PCP d^{-1} in a horizontal-flow anaerobic immobilized biomass reactor. Beaudet et al. (1997) achieved complete dechlorination of PCP in an anaerobic FFBR fed with up to 68 μM d^{-1} of PCP extracted from contaminated wood chips. Among the high-performance anaerobic systems, UASB reactors have shown a better ability for treating higher PCP loads than FF bioreactors (Duff et al., 1995).

Under anaerobic conditions PCP is biodegraded through reductive dechlorination, leading to less-chlorinated chlorophenols and, in some cases, to complete dechlorination, which facilitates subsequent biodegradation (Wu et al., 1993). PCP anaerobic conversion has been mostly carried out under mesophilic conditions, which require around 30% of the energy associated to the biogas generated. However, low and medium strength wastewaters are commonly discharged at low ambient temperatures, including municipal wastewater and a broad variety of industrial wastewater. The treatment at low-temperature by psychrophilic anaerobic bioreactors has recently proved to be feasible for a range of wastewater categories, including the removal of trichloroethylene (Siggins et al., 2011), trichlorophenols (Collins et al., 2005) and nutrients (Ma et al., 2013). These emerging technologies represent a breakthrough for the management of highly polluted wastewaters. The high-performance anaerobic systems, like expanded granular sludge bed (EGSB) reactor, are promising potential solutions where the contact between the biomass and wastewater is enhanced. The high upflow velocity that can be applied in EGSB (4–10 m h^{-1}) provokes effective mixing, allowing improved efficiency (Puyol et al., 2009).

Wastewater treatment by EGSB reactors at low temperatures seems to be particularly promising for regions in which sewage temperatures do not drop below 15 °C (Lettinga et al., 2001) like in subtropical regions, including the Mediterranean countries. Thus, psychrophilic anaerobic digestion represents an attractive alternative and economically sound option for sustainable wastewater management. However a more in-depth knowledge of the response of low-temperature anaerobic systems to highly toxic micropollutants is needed. In this work, the anaerobic decontamination of synthetic wastewater bearing PCP at low temperature by an EGSB reactor is studied with the aim of analyzing the inhibition of methanogenesis caused by PCP which can provide useful information for design.

2.2 METHODS

2.2.1 WASTEWATER COMPOSITION

Synthetic wastewater was prepared by adding the following components (mg L^{-1}): peptone (17.4), yeast extract (52.2), milk powder (116.2), sunflower oil (29), sodium acetate (79.4), starch (122) and urea (91.7) giving a COD around 1.75 g L^{-1}. The fed wastewater was supplemented with 1 mL L^{-1} of the following micronutrients solution (mg L^{-1}): FeCl$_2$·H$_2$O (2000), H$_3$BO$_3$ (50), ZnCl$_2$ (50), CuCl$_2$·H$_2$O (38), MnCl$_2$·4H$_2$O (500), (NH$_4$)6Mo$_7$O$_{24}$·4H$_2$O (50), AlCl$_3$·6H$_2$O (90), CoCl$_2$·6H$_2$O (2000), NiCl$_2$·6H$_2$O (92), Na$_2$SeO·5H$_2$O (162), EDTA (1000), resazurin (0.2), sulphuric acid 36% (1 mL L^{-1}). PCP was incorporated at 10 mg L^{-1}. Sodium bicarbonate was added as buffer and alkalinity source at 1 g g^{-1} COD.

2.2.2 BIOMASS SOURCE

The anaerobic granular sludge was retrieved from a full scale UASB reactor treating brewery wastewater (Alovera, Guadalajara, Spain). The granules presented an average diameter of 1–2 mm and a specific methanogenic activity (SMA) of 0.17 g COD g^{-1} VSS d^{-1}.

2.2.3 EXPERIMENTAL SETUP OF CONTINUOUS RUNS

A 4.2 L lab-scale EGSB reactor inoculated with 100 g VS L^{-1} of granular sludge was used, whose main characteristics are given elsewhere (Puyol et al., 2009). The reactor was generally operated at a HRT of 1 d, room temperature (17–28 °C) and pH around 7.5. The experimental program included two stages. Firstly, the effect of the upward flow rate on PCP conversion and reactor performance was analyzed within a range of 1–4 m h^{-1}. Then, the stability of the EGSB reactor and its effect on PCP conversion was checked by increasing the urea, oils and PCP loading rates (LR) along the experiment. Each operating period was followed by a recovery stage to restore the steady state, in which 4 g COD L^{-1} d^{-1} of a mixture of sodium acetate, propionate and butirate (1:1:1 w:w), and 4 g glucose L^{-1} d^{-1} were used as carbon sources.

2.2.4 BIODEGRADABILITY AND INHIBITION TESTS

Biodegradability tests were performed by duplicate in 250 mL serum bottles inoculated with 1.5 g VS L^{-1} of active granular sludge. Temperature was fixed at 30 ± 1 °C. The specific methanogenic activity (SMA) of the inoculum was estimated according to (Puyol et al., 2009). The synthetic wastewater was spiked with different concentrations of PCP (0–50 mg L^{-1}). The inhibition of methanogenesis was studied using the aforementioned standard methanogenic medium supplemented with different PCP concentrations (0–50 mg L^{-1}) and 4 g CH_3COONa L^{-1} or 2 g HCOONa L^{-1} for the evaluation of the inhibition affecting to acetoclastic and hydrogenotrophic methanogenesis, respectively. Blanks were conducted for all the biotic tests.

2.2.5 ANALYTICAL METHODS

Chlorophenols were quantified by HPLC/UV at 270 nm (Varian Prostar 330), using a C_{18} column as the stationary phase (Valco Microsorb-MW 250-4.6 C_{18}) and a mixture of acetonitrile and 0.025 M acetic acid in deionized water (60:40 v:v) as mobile phase at 0.8 mL min^{-1}. Column temperature was 40 °C. Acetate was quantified by HPLC coupled with a refraction index detector (Varian Prostar 800RI) using sulfonated polystyrene resin in the protonated form (67H type) as stationary phase (Varian Metacarb 67H 300-6.5) and 0.25 mM sulphuric acid in deionized water as the mobile phase at 0.8 mL min^{-1}. Column temperature was 65 °C.

TOC was measured by an OI Analytical Model 1010 TOC apparatus. Analyses of COD, total and volatile suspended solids (TSS and VSS) and biomass concentration (as volatile solids, VS) were performed according to the APHA Standard Methods (APHA, 2005).

2.2.6 KINETIC AND MODEL STUDY

2.2.6.1 INHIBITION OF METHANOGENESIS

Modelization of methanogenesis inhibition was carried out assuming that biomass growth is negligible and taking into account that there is no competence for methanogenic substrates, so inhibition can be exclusively attributed to the presence of PCP. The methane production was fitted to different inhibition models derived from a pseudo-Monod model Eq. (1):

$$\frac{dM}{dt} = V_{max} \cdot \frac{(M_{max} - M)}{K_M + (M_{max} - M)} \tag{1}$$

where V_{max} is the maximum specific methanogenic rate (g CH_4-COD g^{-1} VS d^{-1}), M is the specific methane production, M_{max} is the maximum specific methane production and K_M the saturation constant for methanogenesis (g CH_4-COD g^{-1} VS)

2.2.6.2 INHIBITION OF PCP REMOVAL RATE

The inhibition effect over the PCP removal rates was described by the Eq. (2) assuming no biomass growth:

$$\frac{d[PCP]}{dt} = \frac{k \cdot [PCP]}{1 + \left(\frac{[PCP]}{K_i}\right)^n} \tag{2}$$

where k is the apparent first order rate constant (d^{-1}), K_i is the inhibition constant (mg PCP L^{-1}) and n is the inhibition order (dimensionless).

2.2.6.3 INHIBITION OF COD CONSUMPTION RATE

The inhibitory effect of PCP on COD consumption rate was modeled with a pseudo-Monod model equation Eq. (3):

$$\frac{dCOD}{dt} = V_{max} \frac{COD - COD_n}{K_s + COD - COD_n} \tag{3}$$

where V_{max} is the maximum COD consumption rate (mg COD L^{-1} d^{-1}), COD_n is the non-biodegradable COD fraction and K_s is the saturation constant (mg COD L^{-1}).

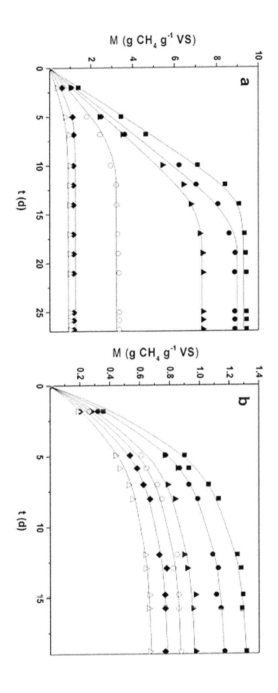

FIGURE 1: Time-evolution of methane production in acetoclastic (a) and hydrogenotrophic (b) experiments at PCP initial concentrations of 0 (■), 1 (●), 5 (triangle symbnol), 10 (○), 25 (◆), and 50 (Δ) mg L⁻¹. Experimental values (symbols) and model predictions (lines).

2.3 RESULTS AND DISCUSSION

2.3.1 METHANOGENIC INHIBITION

Fig. 1 shows methane production from acetoclastic (1a) and hydrogeno-trophic (1b) methanogenesis at different PCP initial concentrations. The maximum methane production significantly diminished in both cases at increasing PCP concentrations, suggesting the occurrence of irreversible inhibition phenomena. The effect on acetoclastic methanogenesis is more accused than for hydrogenotrophic. This fact is in agreement with the EC_{50} values obtained (3 and 20 mg PCP L^{-1} for acetoclastic and hydrogeno-trophic methanogenesis, respectively). It is known that the presence of PCP inhibits all VFA-degrading population, being specially affected the propionate degraders (Wu et al., 1993). However, the nature of the inhibi-tion effect of PCP over methanogens is not clear so far. In all cases the specific acetoclastic methanogenesis activity was higher than the hydroge-notrophic, which suggests that acetoclastic methanogens prevailed over hydrogenotrophic in the granular sludge.

A kinetic study of the methanogenesis inhibition was carried out to know the different behavior of acetoclastic and hydrogenotrophic metha-nogens. In the case of acetoclastic methanogenesis, a remnant production of methane was observed at PCP concentrations from 25 mg L^{-1} onwards, suggesting that there was an early methane production which was not af-fected by PCP. Since M_{max} is clearly affected by the PCP concentration, Eq. (1) becomes Eq. (4):

$$\frac{dM}{dt} = V_{max} \cdot \frac{a \cdot \left(\dfrac{M_{max}}{1 + \left(\dfrac{[PCP]}{k_i}\right)^n} - M\right) + (1 - a) \cdot (M_{max} - M)}{K_M + a \cdot \left(\dfrac{M_{max}}{1 + \left(\dfrac{[PCP]}{k_i}\right)^n} - M\right) + (1 - a) \cdot (M_{max} - M)}$$

$$(4)$$

where a is the fraction of acetoclastic methane production affected by PCP, K_i is the inhibition constant (mg PCP L^{-1}) and n is the inhibition order (dimensionless).

Regarding to the hydrogenotrophic methanogenesis, no remnant methane production was observed, and PCP concentration only affected to M_{max}. Thus, Eq. (1) becomes Eq. (5):

$$\frac{dM}{dt} = V_{max} \cdot \frac{M_{max} \cdot 1 + \left(\frac{[PCP]}{k_i}\right)^n - M}{K_M + \frac{M_{max}}{1 + \left(\frac{[PCP]}{k_i}\right)^n} - M}$$

(5)

Integration of Eq. (4) and (5) was accomplished by the Episode numerical method for Stiff systems with the initial conditions t = 0; M = 0. Experimental data were fitted to the equations proposed by means of a non-linear least squares minimization of the error using a simplex algorithm followed by a Powell minimization algorithm (Micromath® Scientist 3.0). The values of the fitting parameters are listed in Table 2 and the fitting curves are depicted in Fig. 1. High values of the correlation coefficient were obtained and the curves of Fig. 1 show the validity of the models, most in particular in the case of acetoclastic methanogenesis. The inhibition order is significantly lower for hydrogenotrophic methanogenesis (0.17 vs 3.10) consistently with the lower inhibitory effect observed.

TABLE 2: Values of fitting parameters from Eqs. [4] and [5] for methane production from acetoclastic and hydrogenotrophic metanogenesis in presence of PCP.

Methanogenesis	Vmax (g CH$_4$-COD g^{-1} VS d^{-1})	KM (g CH$_4$-COD g^{-1})	Mmax (g CH$_4$-COD g^{-1})	Ki (mg PCP L^{-1})	n	a	R^2
Acetoclastic	1.115 ± 0.034	0.47 ± 0.21	13.901 ± 0.076	7.3 ± 0.08	3.10 ± 0.10	0.905 ± 0.005	0.999
Hydrogenotrophic	0.251 ± 0.014	0.33 ± 0.18	2.004 ± 0.042	0.942 ± 0.326	0.17 ± 0.01	–	0.996

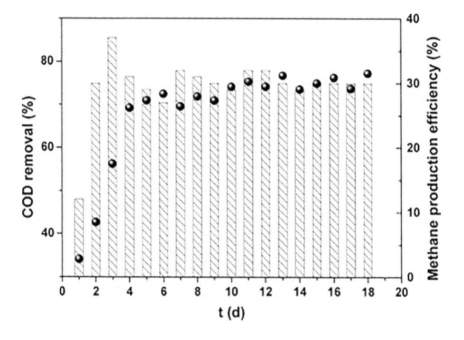

FIGURE 2: COD removal (circles) and methane production efficiency (bars) during the start-up of the EGSB reactor.

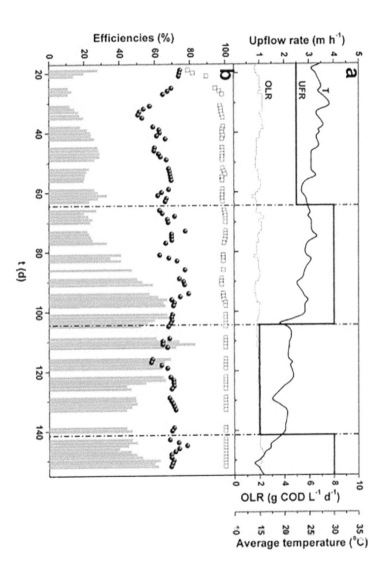

FIGURE 3: Operating conditions (a) and reactor performance (b) in the long-term continuous experiment at different upward flow rate (UFR), PCP conversion (squares), COD consumption (spheres) and methane production (bars) efficiencies. Vertical dash-dot lines indicate changes in operating conditions.

2.3.2 REACTOR PERFORMANCE

2.3.2.1 START-UP

The EGSB reactor was started-up in the absence of PCP using an HRT of 1 d and an upward flow rate of 2.5 m h^{-1}. Around the fourth day quite stable COD consumption and methane production efficiency ($\approx$75 and 30%, respectively) were achieved (Fig. 2). That steady operation was maintained for two weeks where the rector showed a good physiological activity and after the 18th day PCP was incorporated into the feed.

2.3.2.2 EFFECT OF UPWARD FLOW RATE

Fig. 3 summarizes the operating conditions (3a) and reactor performance (3b). Changes in the upward flow rate showed a remarkable effect on methane production. A methane production efficiency of 60% was reached operating at 4 m h^{-1} and this upward flow rate was selected for the rest of the experiment. In spite of the fact that fluctuations of temperature (room conditions) occurred along the experiment (Fig. 3a), the PCP conversion and COD consumption efficiencies remained almost unaltered. With regard to the methanogenesis efficiency around 50–60% was in general observed beyond about the 90th day of operation up to the end of the experiment (170 days).

A COD removal efficiency around 75% was reached after the stabilization of the reactor during the start-up period (Fig. 2), which decreased dramatically when PCP was added at the 18th day. The rapid increase of the PCP concentration could cause the accumulation of more toxic intermediates (i.e., trichlorophenols). In some cases, this phenomenon has led to the destabilization of the bioreactor (Tartakovsky et al., 2001). Nevertheless, gradual exposure to PCP allowed the adaption of the anaerobic consortium, specially the xenobiotic-sensitive methanogens (Angelidaki et al., 2003). The toxic effects caused by PCP could be mitigated by available carbon sources in the anaerobic medium (Ye and Shen, 2004), which would allow the effective conversion of PCP. Sugars and alcohols have been widely used as cosubstrates to favor PCP dechlorination (Puyol et al., 2009). In this work, the addition of species related with municipal wastewater components allowed treating efficiently PCP at 16.5 mg PCP L^{-1} d^{-1} loading rate.

2.3.2.3 STABILITY ANALYSIS

Fig. 4a shows the operating conditions of the EGSB reactor along the stability test, where OLR was modified. These changes were accomplished by duplicating the corresponding LR maintaining the HRT constant in all the cases except for the PCP-LR increment, where a reduction of the HRT by a half was applied. The reactor performance is shown in Fig. 4b.

The step increase of urea concentration in the inlet stream did not provoke a significant effect on the PCP and COD removal efficiencies. Eventual variations of COD removal during this stage can be related with changes in the reactor temperature. In contrast, the increase of sunflower oil LR caused a severe drop of the COD removal efficiency without affecting to PCP conversion but accompanied by a consistent decrease of methanogenesis efficiency. This fact could be explained by the occurrence of unspecific inhibition phenomena, which could affect the organic matter consumption while remaining the reductive dechlorination activity unaltered.

Both high LR of urea and sunflower oil showed a detrimental effect on the methanogenesis efficiency. However, the performance of the reactor was subsequently recovered by adding easily biodegradable components, which suggest that the inhibition effect was reversible in both cases. Although synthetic wastewater composition in this work was really complex, a timely acclimation period was necessary to recover the degradation efficiencies when the PCP-LR was duplicated (33 mg PCP L^{-1} d^{-1}), mainly due to the reduction of the activity of acetoclastic methanogens (Shen et al., 2005).

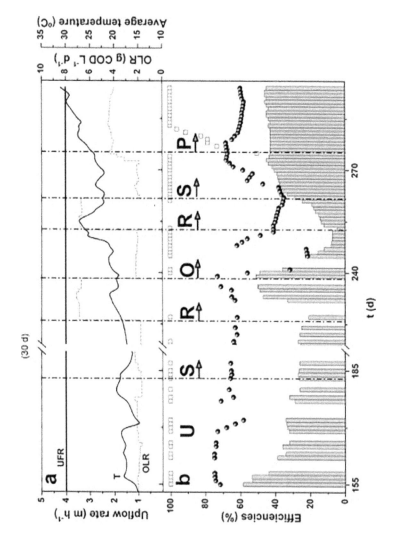

FIGURE 4: Operating conditions (a) and reactor performance (b) under step inputs of different compounds present in wastewater. S: standard EGSB operation mode at 2 g COD L⁻¹ d⁻¹ and 4 m h⁻¹; R: recovery period; O: sunflower oil peak; P: PCP peak. Rest of nomenclature as in Fig. 3.

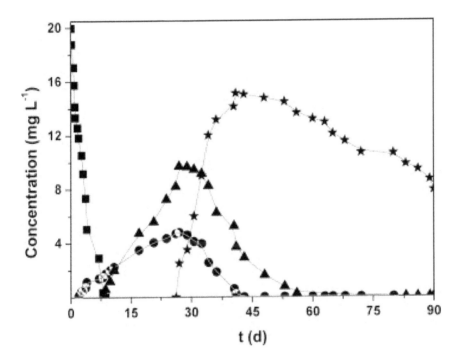

FIGURE 5: Time-evolution of PCP and its chlorinated intermediates: PCP (squares), 235TCP (circles), 23DCP (triangles) and 3CP (stars). [PCP]0 = 20 mg L^{-1}.

2.3.3 KINETIC ANALYSIS

Biodegradability tests were performed using adapted granular sludge retrieved from the EGSB reactor after the long-term continuous experiment of Fig. 3. The time-evolution of PCP and its chlorinated intermediates is reported in Fig. 5. As can be seen, the initial dechlorination of PCP occurred at ortho and para positions, generating 2,3,5-TCP, which was dechlorinated to 3,5-DCP at ortho position, and then to 3-CP at meta position, which was identified as a refractory compound. However, a sharp drop of PCP was registered during the first hours, which can be related to the occurrence of bioaccumulation (Puyol et al., 2011). Although 3-CP, the main PCP degradation intermediate, was still present in the resulting effluent, a PCP removal rate three times higher than the reported in previous works was achieved in spite of the lower temperature of our experiments.

After the bioaccumulation period, initial PCP removal rates were obtained for different starting PCP concentrations. It was observed that the initial PCP removal rate increase with PCP concentration up to around 10 mg L^{-1}, followed by a clear reduction beyond that concentration. The experimental data (not shown) were fitted to Eq. (2) using the Levenbert-Marquart algorithm (Microcal Origin® 7.5). The following values were obtained for the fitting parameters: $k = 0.62 \pm 0.11$, $K_i = 10.02 \pm 1.75$, $n = 2.75 \pm 0.48$ and $R^2 = 0.95$, which shows the validity of Eq. [2].

Fig. 6a shows the time-evolution of COD treating different PCP concentrations. A pseudo-Monod model equation Eq. (3) was used to describe the inhibitory effect of PCP on COD consumption rate. The values of K_s, V_{max} and COD_n, obtained for different initial PCP concentration (data not shown), indicate that PCP exerts an inhibitory effect which can be described with the inclusion of a power term associated to each parameter. Considering the same inhibitory effect on V_{max} and K_s (Haldane-type or uncompetitive inhibition), Eq. (3) becomes Eq. (6):

$$\frac{dCOD}{dt} = V_{max} \cdot \frac{COD - COD_n \cdot \left(1 + \left(\frac{Ki_1}{[PCP]}\right)^n\right)}{K_s + COD - COD_n \cdot \left(1 + \left(\frac{Ki_1}{[PCP]}\right)^n\right) + \left(COD - COD_n \cdot \left(1 + \left(\frac{Ki_1}{[PCP]}\right)^n\right)\right) \cdot \left(\frac{Ki_2}{[PCP]}\right)^m} \tag{6}$$

where Ki_1 and Ki_2 (mg PCP L^{-1}) are the uncompetitive and the suicide inhibition constants, respectively, n and m are the inhibition orders (dimensionless). Integration of Eq. (6) was performed as described for the inhibition tests, with the initial condition $t = 0$; COD = $CODo$, considering PCP as constant and equal to initial PCP concentration, since the PCP inhibitory effect of PCP can be considered the same at all the concentrations tested. The fitting curves are included in Fig. 6a which shows the validity of Eq. [6]. The following values were obtained for the fitting parameters: V_{max} = 549 ± 50 mg COD ^{-1}l d^{-1}, Ks = 647 ± 174, COD_n = 500 ± 26 mg L^{-1}, Ki_1 = 7.4 ± 1.7 mg PCP L^{-1}, Ki_2 = 7.3 ± 1.0 mg PCP L^{-1}, n = 0.496 ± 0.031 and m = 0.823 ± 0.053 (R^2 = 0.999).

The value of COD_n (500 mg L^{-1}) for a COD initial concentration around 2500 mg L^{-1} is consistent with the COD efficiency achieved in the EGSB reactor to around 80% at the most. This fact supports the idea that a fraction of the COD can be considered as hardly biodegradable in presence of PCP under the operating conditions tested. However, the values of COD_n increased with PCP concentration indicating a severe toxic effect on the biodegradation capacity of the sludge. This could be explained by a selective detrimental effect over a specific group of microorganisms, which leads to the accumulation of some byproducts of anaerobic digestion of the low-strength synthetic wastewater used. Taking into account that acetate was the major organic product detected after the biodegradability tests, acetoclastic methanogens and other acetate-degraders could be considered as the main sensitive microbial groups affected by PCP.

As can be observed in Fig. 6b, methanogenesis was also critically affected by PCP, diminishing both the maximum methane production and the methanogenic rate as the OCO concentration was increased. However, a basal methane production was observed regardless the presence of PCP since at the highest PCP concentrations tested (20 and 50 mg L^{-1}), methane production was almost equal. The methanogenesis rate was described by a simplified form of Eq. (1) to a pseudo-first order equation, commonly known as the Roediger model (Puyol et al., 2011). This model was modified by including both the inhibition effects detected (the decrease of the maximum methane production and methanogenesis rate) and the basal methane production, yielding Eq. (7):

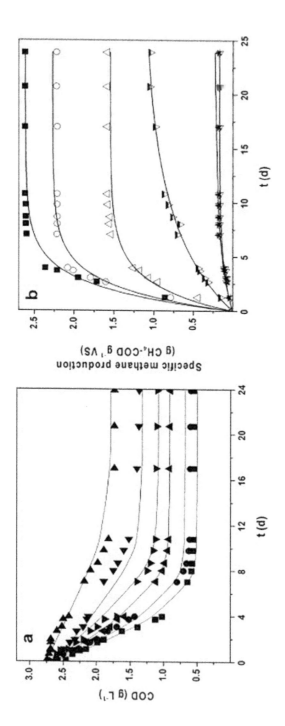

FIGURE 6: Time-evolution of COD (a) and specific methane production (b) at different PCP starting concentrations (0–50 mg L^{-1}). Experimental data (symbols) and model fittings (continuous lines).

$$\frac{dM}{dt} = k_b \cdot k \cdot \left(\frac{M_{max}}{1 + \left(\frac{Ki_1}{[PCP]}\right)^n} - M\right) + \frac{k_M \cdot (1 - k) \cdot \left(\frac{M_{max}}{1 + \left(\frac{Ki_1}{[PCP]}\right)^n} - M\right)}{1 + \left(\frac{Ki_2}{[PCP]}\right)^m}$$

(7)

where k_b and k_M are the apparent pseudo-first order rate constants (d^{-1}) for the basal and PCP-affected methane productions, respectively, Ki_1 and Ki_2 are the inhibition constants for the detected inhibition on M_{max} and kM (mg PCP L^{-1}), and n and m are the inhibition orders. Eq. (7) was integrated by the aforementioned method with the initial condition $t = 0$; $M = 0$, as-suming that the inhibition effect caused by PCP remains invariable. The fitting curves are depicted in Fig. 6b. The fitting values for the parameter were: $M_{max} = 2.446 \pm 0.032$ g CH_4-COD g^{-1} VSS, $Ki_1 = 6.495 \pm 0.427$ mg PCP L^{-1}, n = 0.970 ± 0.071, kM = 0.478 ± 0.022 d^{-1}, $Ki_2 = 7.827 \pm 0.560$ mg PCP L^{-1}, m = 5.055 ± 1.065 and $k_b = 0.282 \pm 0.024$ d^{-1} ($R^2 = 0.990$). Methanogens, especially acetate-consumers, seem to be the main popula-tion affected by the PCP toxicity despite the acclimation of the sludge. Nevertheless, the values of the inhibition parameters obtained when using adapted biomass indicate that the resistance to PCP was enhanced along the continuous experiment.

2.4 CONCLUSIONS

Low-strength wastewater bearing PCP can be treated in an EGSB reac-tor operated at 4 m h^{-1} upward flow rate and PCP loads up to around 16 mg PCP L^{-1} d^{-1}, being the anaerobic activity barely affected at psychro-mesophilic temperatures. Changes in loading rate and step inputs of urea and sunflower oil, caused a certain reversible effect over methanogenesis and COD consumption, remaining the PCP conversion almost unaltered.

Increasing PCP concentration caused a decrease of COD removal and an irreversible inhibitory effect on methanogenesis, specially over the acetoclastic methanogens. The results allow concluding that anaerobic treatment in EGSB reactor can be a promising solution for the treatment of PCP-containing wastewaters.

REFERENCES

1. American Public Health Association. 2005. Standard methods for the examination of water and wastewater. APHA-AWWA-WEF, Washington, D.C.
2. I. Angelidaki, L. Ellegaard, B. Ahring. Applications of the anaerobic digestion process. Biomethanation II, vol. 82Springer, Berlin (2003), pp. 1–33
3. E.A. Baraldi, M.H.R.Z. Damianovic, G.P. Manfio, E. Foresti, R.F. Vazoller. Performance of a horizontal-flow anaerobic immobilized biomass (HAIB) reactor and dynamics of the microbial community during degradation of pentachlorophenol (PCP). Anaerobe, 14 (2008), pp. 268–274
4. R. Beaudet, G. McSween, F. Lépine, S. Milot, J.G. Bisaillon. Anaerobic biodegradation of pentachlorophenol in a liquor obtained after extraction of contaminated chips and wood powder. J. Appl. Microbiol., 82 (1997), pp. 186–190
5. Y. Chen, J.J. Cheng, K.S. Creamer. Inhibition of anaerobic digestion process: a review. Bioresour. Technol., 99 (2008), pp. 4044–4064
6. N. Christiansen, B. Ahring. Introduction of ade novo bioremediation activity into anaerobic granular sludge using the dechlorinating bacterium DCB-2. Antonie van Leeuwenhoek, 69 (1996), pp. 61–66
7. G. Collins, C. Foy, S. McHugh, V. O'Flaherty. Anaerobic treatment of 2,4,6-trichlorophenol in an expanded granular sludge bed-anaerobic filter (EGSB-AF) bioreactor at 15 °C. FEMS Microbiol. Ecol., 53 (2005), pp. 167–178
8. G.S. Cooper, S. Jones. Pentachlorophenol and cancer risk: focusing the lens on specific chlorophenols and contaminants. Environ. Health Perspect., 116 (2008), pp. 1001–1008
9. M.H. Damianovic, E.M. Moraes, M. Zaiat, E. Foresti. Pentachlorophenol (PCP) dechlorination in horizontal-flow anaerobic immobilized biomass (HAIB) reactors. Bioresour. Technol., 100 (2009), pp. 4361–4367
10. T.J. Downs, E. Cifuentes-Garcia, I.M. Suffet. Risk screening for exposure to groundwater pollution in a wastewater irrigation district of the Mexico City Region. Environ. Health Perspect., 107 (1999), pp. 553–561
11. S.J.B. Duff, K.J. Kennedy, A.J. Brady. Treatment of dilute phenol/PCP wastewaters using the upflow anaerobic sludge blanket (UASB) reactor. Water Res., 29 (1995), pp. 645–651
12. T. Essam, M.A. Amin, O. El Tayeb, B. Mattiasson, B. Guieysse. Sequential photochemical-biological degradation of chlorophenols. Chemosphere, 66 (2007), pp. 2201–2209

13. M. Farhadian, D. Duchez, C. Vachelard, C. Larroche. Monoaromatics removal from polluted water through bioreactors – a review. Water Res., 42 (2008), pp. 1325–1341
14. J. Gasperi, S. Garnaud, V. Rocher, R. Moilleron. Priority pollutants in wastewater and combined sewer overflow. Sci. Total Environ., 407 (2008), pp. 263–272
15. P. Juteau, R. Beaudet, G. McSween, F. Lépine, S. Milot, J.G. Bisaillon. Anaerobic biodegradation of pentachlorophenol by a methanogenic consortium. Appl. Microbiol. Biotechnol., 44 (1995), pp. 218–224
16. A.P. Khodadoust, J.A. Wagner, M.T. Suidan, R.C. Brenner. Anaerobic treatment of PCP in fluidized-bed GAC bioreactors. Water Res., 31 (1997), pp. 1776–1786
17. K.M. Koran, M.T. Suidan, A.P. Khodadoust, G.A. Sorial, R.C. Brenner. Effectiveness of an anaerobic granular activated carbon fluidized-bed bioreactor to treat soil wash fluids: a proposed strategy for remediating PCP/PAH contaminated soils. Water Res., 35 (2001), pp. 2363–2370
18. M. Lanthier, P. Juteau, F. Lépine, R. Beaudet, R. Villemur. Desulfitobacterium hafniense is present in a high proportion within the biofilms of a high-performance pentachlorophenol-degrading, methanogenic fixed-film reactor. Appl. Environ. Microbiol., 71 (2005), pp. 1058–1065
19. R.J. Letcher, W.A. Gebbink, C. Sonne, E.W. Born, M.A. McKinney, R. Dietz. Bioaccumulation and biotransformation of brominated and chlorinated contaminants and their metabolites in ringed seals (Pusa hispida) and polar bears (Ursus maritimus) from East Greenland. Environ. Int., 35 (2009), pp. 1118–1124
20. G. Lettinga, J.B.v. Lier, J.C.L.v. Buuren, G. Zeeman. Sustainable development in pollution control and the role of anaerobic treatment. Water Sci. Technol., 44 (6) (2001), pp. 181–188
21. B. Ma, Y. Peng, S. Zhang, J. Wang, Y. Gan, J. Chang, S. Wang, S. Wang, G. Zhu. Performance of anammox UASB reactor treating low strength wastewater under moderate and low temperatures. Bioresour. Technol., 129 (2013), pp. 606–611
22. H. Mohn, J.A. Puhakka, J.F. Ferguson. Effects of electron donors on degradation of pentachlorophenol in a methanogenic fluidized bed reactor. Environ. Technol., 20 (1999), pp. 909–920
23. M.A. Montenegro, E.M. Moraes, H.M. Soares, R.F. Vazoller. Hybrid reactor performance in pentachlorophenol (pcp) removal by anaerobic granules. Water Sci. Technol., 44 (2001), pp. 137–144
24. C.H. Mun, J. He, W.J. Ng. Pentachlorophenol dechlorination by an acidogenic sludge. Water Res., 42 (2008), pp. 3789–3798
25. M. Nielsen, T. Pettersen, D. Miljøstyrelsen. Genanvendelse af gråt spildevand på campingpladser – fase 2 og 3. Miljøstyrelsen (2005)
26. D. Puyol, A.F. Mohedano, J.L. Sanz, J.J. Rodríguez. Comparison of UASB and EGSB performance on the anaerobic biodegradation of 2,4-dichlorophenol. Chemosphere, 76 (2009), pp. 1192–1198
27. D. Puyol, A.F. Mohedano, J.J. Rodriguez, J.L. Sanz. Effect of 2,4,6-trichlorophenol on the microbial activity of adapted anaerobic granular sludge bioaugmented with desulfitobacterium strains. New Biotechnol., 29 (2011), pp. 79–89
28. D.V. Savant, R. Abdul-Rahman, D.R. Ranade. Anaerobic degradation of adsorbable organic halides (AOX) from pulp and paper industry wastewater. Bioresour. Technol., 97 (2006), pp. 1092–1104

29. D.-S. Shen, X.-W. Liu, H.-J. Feng. Effect of easily degradable substrate on anaerobic degradation of pentachlorophenol in an upflow anaerobic sludge blanket (UASB) reactor. J. Hazard. Mater., 119 (2005), pp. 239–243

30. D.-S. Shen, R. He, X.-W. Liu, Y. Long. Effect of pentachlorophenol and chemical oxygen demand mass concentrations in influent on operational behaviors of upflow anaerobic sludge blanket (UASB) reactor. J. Hazard. Mater., 136 (2006), pp. 645–653

31. A. Siggins, A.M. Enright, V. ÓFlaherty. Temperature dependent (37–15 °C) anaerobic digestion of a trichloroethylene-contaminated wastewater. Bioresour. Technol., 102 (2011), pp. 7645–7656

32. R.R. Singhania, G. Christophe, G. Perchet, J. Troquet, C. Larroche. Immersed membrane bioreactors: an overview with special emphasis on anaerobic bioprocesses. Bioresour. Technol., 122 (2012), pp. 171–180

33. B. Tartakovsky, M.F. Manuel, D. Beaumier, C.W. Greer, S.R. Guiot. Enhanced selection of an anaerobic pentachlorophenol-degrading consortium. Biotechnol. Bioeng., 73 (2001), pp. 476–483

34. W.M. Wu, L. Bhatnagar, J.G. Zeikus. Performance of anaerobic granules for degradation of pentachlorophenol. Appl. Environ. Microbiol., 59 (1993), pp. 389–397

35. F.-X. Ye, D.-S. Shen. Acclimation of anaerobic sludge degrading chlorophenols and the biodegradation kinetics during acclimation period. Chemosphere, 54 (2004), pp. 1573–1580

The Behavior of Organic Phosphorus Under Non-Point Source Wastewater in the Presence of Phototrophic Periphyton

HAIYING LU, LINZHANG YANG, SHANQING ZHANG, AND YONGHONG WU

3.1 INTRODUCTION

The discharge of excessive phosphorus into isolated water bodies will accelerate the eutrophication process. These water bodies such as lakes and dams suffer from severe water quality problems that are closely linked with the excessive phosphorus inputs from various pollution sources. This happens in both developed and developing countries [1], [2]. Thus, the removal measures of phosphorus from these water bodies, especially biological methods based on biofilms, have currently been a subject of great concern [3].

Organic phosphorus (P_{org}) commonly includes nucleic acids, phospholipids, inositol phosphates, phosphoamides, phosphoproteins, sugar phosphates, amino phosphoric acids and organic condensed phosphorus species. It is often at least as abundant as (sometimes great excess of) inorganic phosphorus (P_{inorg}) in natural water bodies and sediments [4]. Previous studies show that soluble P_{org} in water system, especially in lakes, often exceeded that of orthophosphate and accounted for 50%–90% of total phosphorus [5], [6]. In aquatic systems, the role of P_{org} has largely been underestimated not only because of its complexity in composition and structure [7], but also due to the limitation in analytical methods and techniques. As a result, P_{org} has usually been grouped in the "non-reactive" and "non-bioavailable" component of total phosphorus (P_{total}) [4]. However, there is strong evidence that some organisms such as algae and bacteria are adapted to P_{org} via enzymatic hydrolysis and/or bacterial decomposition [8]–[10]. As a result, the importance of P_{org} is not widely recognized as a potentially large pool of bioavailable phosphorus, and its influences on phosphorus cycling and eutrophication of aquatic ecosystem are inevitably ignored among organic phosphorus species. Most importantly, P_{org} is typically not susceptible to the traditional removal technologies for the inorganic phosphorus [11], which may be due to its complicated species and chemical dynamics.

Phototrophic periphyton is mainly composed of multilayered consortia of photoautotrophs (e.g., cyanobacteria and microalgae) and heterotrophs (e.g. bacteria, fungi and protozoa), which is dominated by photoautotrophic microorganisms. These multilayer constructions are embedded in a common extracellular polymeric substance (EPS), secreted by the community, which mediates the adhesion of phototrophs and heterotrophs as well as gas and nutrient fluxes [12]. The periphyton is ubiquitous in aquatic environments and performs numerous important environmental functions such as nutrients cycling and self-purification of aquatic ecosystems [13], [14]. It has been proven that the periphyton has a high affinity for inorganic phosphorus and can act as an important potential sink for phosphorus in wetlands [15]. Thus, the periphyton has subsequently been developed to remove inorganic phosphorus from wastewaters due to its cost-effectiveness, easy-harvesting, high-effectiveness and environment-friendly advantages [16]. However, information about organic phosphorus utilization and removal by the periphyton such as kinetic analysis is still very limited. Although ATP represented by organic phosphorus had been investigated [17], the detailed removal mechanism and transformation as well as removal kinetics are still not clear.

Also, most current studies about phosphorus removal methods or technologies were focused on purely inorganic phosphorus or a specific type of phosphorus-based contaminants such as organophosphorus pesticides [18]. Moreover, it was postulated previously that the periphyton was capable of transforming P_{org} to P_{inorg} because of high phosphatase activities [19], which could introduce confusion between removal and conversion of P_{org} due to that P_{org} is traditonally calculated as the difference between P_{total} and P_{inorg}. Therefore, a comprehensive investigation on P_{org} conversion and removal process by the periphyton ubiquitously distributed in natural waters will not only develop a potential technology for P_{org} removal from high-organic waters such as animal wastes, but also provide strong evidence to fully understand the phosphorus biogeochemical cycling of aquatic ecosystem that contain the periphyton or similar microbial aggregates.

In this work, we attempt to remove P_{org} from non-point source wastewater using the periphyton. The main objectives of this study were to (i) quantify the conversion and removal kinetic processes of P_{org} in the presence of the periphyton; (ii) evaluate the removal mechanism of P_{org} by the periphyton; (iii) explore the influence of environmental conditions to P_{org} removal by the periphyton.

3.2 MATERIALS AND METHODS

3.2.1 ETHICS STATEMENT

The study was not involved in any endangered or protected species. The investigation was permitted by Xuanwu Hu lake Management Committee, which is a public and benefit organ.

3.2.2 PHOTOTROPHIC PERIPHYTON CULTURE

The biofilm substrate—Industrial Soft Carriers (Diameter 12 cm and length 55 cm, Jineng environmental protection company of Yixing, China) was used for in situ collecting and culturing of periphyton biofilms from Xuanwu Lake, East China. During the experiment, the substrates were submerged into the lake water (total nitrogen: 1.90 mg L^{-1}, total phosphorus: 0.1 mg L^{-1}, pH: 7.8, ammonia: 0.53 mg L^{-1}; nitrate: 0.73 mg L^{-1}); and the microorganisms in the hypereutrophic water as inoculums attached on the substrate surfaces and formed periphyton biofilms. Once the biofilms was covered on the substrate surface, the periphytons with their substrates were taken out for indoor culture.

The indoor culture of the periphyton was conducted in glass tanks (each tank: 100 cm length, 100 cm width, and 60 cm height). Firstly, the tanks were sterilized using a 95% alcohol solution and rinsed with water. Then, the collected periphyton along with their substrates were submerged into the glass tanks filled with simulated artificial wastewater [composed of macro nutrient (20 mg L^{-1} $NaCO_3$, 150 mg L^{-1} $NaNO_3$, 40 mg L^{-1} K_2HPO_4, 75 mg L^{-1} $MgSO_4 \cdot 7H_2O$, 36 mg L^{-1} $CaCl_2 \cdot 2H_2O$) and micro nutrient (2.86 mg L^{-1} H_3BO_4, 1.81 mg L^{-1} $MnCl_2 \cdot 4H_2O$, 0.22 mg L^{-1} $ZnSO_4$, 0.39 mg L^{-1} Na_2MoO_4, 0.079 mg L^{-1} $CuSO_4 \cdot 5H_2O$, 4.94 mg L^{-1} $Co(NO_3)_2 \cdot 6H_2O$) as well as organic matters (6 mg L^{-1} citric acid and ammonium ferric citrate)]. To avoid the influence of climatic condition on the periphyton growth, the glass tanks were kept in a greenhouse with air temperature at 25–30°C. When dense the periphyton was formed (the thickness of the periphyton exceeded 5 mm) after 60 days, it was collected for the following experiments.

3.2.3 CHARACTERIZATION OF THE PERIPHYTON

The morphology of the periphyton was observed with Optical Microscopy (OM), Scanning Electron Microscope (SEM) and Zeiss Confocal Laser Scanning Microscope (CLSM). The microbial diversity of phototrophic periphyton was investigated using the method of Biolog™ ECO Microplates [20]. Briefly, 1 g of the periphyton (wet weight) were peeled off and cleaned under sterile conditions, and 150 µL aliquots were added into each well of every Biolog™ ECO Microplate, which was incubated at 25°C and color development (590 nm) was evaluated using a Biolog Microplate Reader every 12 h for seven days (168 h). Based on the pre-experiment, the ratio of dry to wet weight of the periphyton is 0.0532 ± 0.0085 (average $\pm$ SD, n = 10), then 5% was selected as standards for dry weight calculation in all experiments subsequently.

3.2.4 P_{ORG} STOCK PREPARATION

Previous studies found that ATP was an effective substrate for tracing organic phosphorus dynamics in phytoplankton and periphyton [17], [21]. Thus, P_{org} stock solution (100 mg P L^{-1}) was prepared by dissolving 0.6505 g ATP (disodium adenosine triphosphate, $C_{10}H_{14}O_{13}N_5P_3Na_2 \cdot 3H_2O$, sigma) into 1 L distilled water. All P_{org} concentrations used in experiments were diluted with the ATP stock.

3.2.5 P_{ORG} CONVERSION EXPERIMENT

The conversion kinetic experiments of P_{org} were conducted in 250-mL flasks that contained 0 (control), 0.05, 0.1, and 0.2 g of the periphyton biomass in an incubator with an initial P_{org} concentration (C_0) of about 20 mg P L^{-1} (conditions: light intensity = 12000 Lux, temperature = 25°C). The total phosphorus (P_{total}) and inorganic phosphorus (P_{inorg}) concentrations in solution were determined after 1, 4, 8, 12, 24, 36, and 48 hours, respectively.

To identify whether the phosphatase responsible for P_{org} conversion process in the presence of the periphyton, the method based on substrate para-nitrophenyl phosphate (pNPP) was used for phosphatase activities determination of the periphyton.

3.2.6 P_{ORG} REMOVAL EXPERIMENT

Five different treatment levels, each with different biomass (0, 0.1, 0.2, 0.4, 0.6 g) of the periphyton, were tested in this work. The periphyton was placed into 250-ml flasks with the artificial non-point source wastewater (without P_{inorg}) of an initial P_{org} concentration at about 13 mg P L^{-1}.

P_{total} and P_{inorg} concentrations in solution were determined in 48 hours. To distinguish which mechanisms (adsorption and assimilation) are responsible for P_{org} removal process, 0.25 g NaN, was added in solution to inhibit microbial activity and then the phosphorus concentrations in solution were determined.

To further evaluate the removal process, batch kinetic assay was conducted in 250-mL flasks under different biomass of phototrophic periphyton (0.2, 0.4 and 0.6 g) and temperature (10, 20, and 30°C). P_{total} and P_{inorg} concentrations in solution were determined after 1, 4, 8, 12, 24, 36, and 48 hours.

3.2.7 SAMPLE ANALYTICAL METHODS

Dry weight (DW) of the periphyton was determined by oven drying samples at 80°C for 72 h. The biomasses of the periphyton used in the whole study (if not explained) were dry weights (g). The morphology of the phototrophic periphyton was characterized by optical microscope (OM), scan electronic microscope (SEM) and confocal laser scanning microscope (CLSM). The phosphatase assay procedure broadly follows that of Ellwood et al [19], Briefly, the periphyton biomasses cultured in different P_{org} concentration (from 10 to 50 mg P L^{-1}) were carefully prepared, divided into similar sized aliquots and placed into 15-mL tubes containing 9.5 ml of artificial non-point source wastewater (without N or P), while the control contained no phototrophic periphyton. The tubes were then placed in a shaking incubator at 25°C for 20 min before the addition of 0.5 ml substrate (final concentration of 0.25 mM). The samples were then incubated for 3 h, after which the assay reaction was terminated by the addition 0.5 ml of 0.5 M NaOH. Finally, the phosphatase was determined using the absorbance of 405 nm wavelength. Biomasses of the periphyton were removed from the solution, rinsed and dried and weighed to an accuracy of 1.0 mg. The P_{total} and P_{inorg} concentrations in solution were determined simultaneously using a flow injection analyzer (SEAL AA3, German).

3.2.8 DATA ANALYSES

Each experiment in this study was conducted in triplicate, and the mean results (± SD) are presented. Statistical analysis was performed using SPSS 19.0, and $p<0.05$ indicated statistical significance. All figures were derived using Origin 8.0 and Excel 2007.

For the Biolog trial, average well color development (AWCD) was calculated according to the following equation:

$$AWCD = \frac{\Sigma(C - R)}{n}$$

(1)

where C is color production within each well, R is the absorbance value of the plate's control well, and n is the number of substrates (n = 31). Shannon index (H) is commonly used to characterize species diversity in a community, which was obtained by the following equation[22]:

$$H = -\sum pi \ln pi$$

(2)

where p_i is the proportion of the relative absorbance value of well i to total plate's wells.

For enzyme activity assay, phosphatase activity (PA) was calculated by calibration curves constructed from p-Nitrophenol (pNP) standards (0 – 0.2 mM) in assay medium, which was expressed as mmol pNP released g^{-1} DW (dry weight) h^{-1} and obtained by the following equation:

$$PA = \frac{C_i}{mt}V$$

(3)

where C_i is the concentration of pNP (mmol L^{-1}), m is the dry weight of phototrophic periphyton (g), t is reaction time (h), and V is volume of solution.

For conversion kinetic study, the amount of P_{org} transformed to P_i (q_c) at time t (0, 1, 2, 4, 8, 12, 24, and 48 h) was obtained based on Eq. 4:

$$q_c = \frac{(C_t - C_0)V}{m}$$

(4)

where C_0 is the initial P_{inorg} concentration (mg L^{-1}), C_t is the concentration of P_{inorg} at time t, V is the volume of solution (L) and m is the dry weight of the periphyton (g).

Pseudo-first-order kinetic (Eq. 5) and Pseudo-second-order kinetic (Eq. 6) models were used to evaluate the conversion process [23]:

$$\log(q_e - q_c) = \log q_e - \frac{k_l}{2.303}t$$

(5)

where q_c and q_e represent the amount of P_{org} transformed to P_{inorg} (mg g^{-1}) at time t and at equilibrium time, respectively. Parameter k_1 represents the adsorption first-order rate constant (min^{-1}) that calculated from the plot of $\log (q_e - q_c)$ against time (t).

$$\frac{t}{q_c} = \frac{1}{k_2 q_e^2} + \frac{t}{q_e}$$

(6)

where k_2 is the pseudo-second-order rate constant (g mg^{-1} h^{-1}). A plot between t/q_c versus t gives the value of the constants k_2 and also q_e (mg g^{-1}) can be calculated.

For P_{org} removal kinetic study, since there is no P_{inorg} in P_{org} solution, the amount of P_{org} removed (q_t) at time t (0, 1, 2, 4, 8, 12, 24, and 48 h) was obtained based on Eq. 7:

$$q_t = \frac{(C_0 - C_t)V}{m} \tag{7}$$

where C_0 is the initial total phosphorus concentration (mg L^{-1}), C_t is the concentration of total phosphorus at time t, V is the volume of solution (L) and m is the dry weight of the periphyton (g).

Pseudo-first-order kinetic and Pseudo-second-order kinetic models (Eq. 5 and 6) were also used to evaluate the removal process. Moreover, adsorption models such as the intra-particle diffusion (Eq. 8) and Arrhenius equation (Eq. 9) were used to further analyze the removal process, which were calculated as follows:

$$q_t = k_{id}t^{0.5} + C \tag{8}$$

where q_t is the amount removed at time t, parameter K_{id} (mg g^{-1} min$^{0.5}$) is the rate constant of intra-particle diffusion.

$$\ln k = \frac{-E_a}{RT} + \ln A \tag{9}$$

where E_a is activation energy, T is the temperature in Kelvin, R is the gas constant (8.314 J mol^{-1} K^{-1}) and A is a constant called the frequency factor. Value of E_a can be determined from the slope of ln k versus T^{-1} plot.

3.3 RESULTS AND DISCUSSION

3.3.1 CHARACTERISTICS OF THE PHOTOTROPHIC PERIPHYTON

It was observed that the periphyton was mainly composed of green algae, diatoms, bacteria, and protozoa, which was dominated by phototrophic algae (Fig.1). These algae with a diameter of about 1.5 um intertwined each other, forming the base matrix for other microorganisms such as bacteria. The Shannon index of the periphyton based on the Diolog analyses was about 3.1 after 7 days, indicating that there were many types of microorganisms living in the periphyton [24].

Micro-structure is a significant determinant in the activity of the biofilm since it plays an important role in the transportation of nutrients and waters [14]. Previous studies indicated biofilms structure was heterogeneous and complex, which contained voids, channels, cavities, pores, and filaments and with cells arranged in clusters or layers [25]. As shown in the CSLM image that the periphyton composed of biomass clusters separated by interstitial voids, which might have considerable consequences on mass transfer inside the biofilms and exchange of substrates and products with the water phase. Because such micro-voids of the periphyton could play many important roles such as interception in nutrients transportation, especially granular nutrients, between sediment and water interface. However, such structure characteristic was obviously influenced by the species arrangement of organisms that composing the biofilms [26]. For example, the micro-voids constructed among complex cells such as algae, bacteria and protozoa may be larger than these constructed by single species. These voids might also provide more micro-spaces or adsorption sites for capturing nutrients,especially for the particulate nutrients such as polyphosphate particles. It may assisted in the understanding of self-purification of aquatic systems that contain the periphyton [14].

3.3.2 P_o CONVERSION PROCESS BY THE PERIPHYTON

The transformation process of P_{org} (ATP) by the periphyton was studied by monitoring the P_{total}, P_{inorg} and q_c over time (Fig. 2). When the initial P_{org} concentration was about 20 mg P L^{-1}, the P_{inorg} concentrations in solution were obviously increased over times from about 0.7 to 6.4, 10.2, and 14.3 mg P L^{-1} under 0.2, 0.4, and 0.6 g L^{-1} of the periphyton content respectively, while the control (no periphyton) showed no significant change in 48 h (P>0.05). These indicated that the periphyton had relatively substantial transformation ability to convert P_{org} to P_{inorg}, which became stronger with the increasing biomass of the periphyton. It is well known that the reaction rate of (conversion rate of P_{org} to P_{inorg}) was directly associated with phosphatase based on the reaction equation [ATP + enzyme $\rightarrow$ ADP + P_i + energy]. This means that the periphyton produced phosphates, which is beneficial to the P_{org} conversion reaction. Moreover, the content of phosphates increased with the increasing biomass of periphyton.

To quantify the transformation capacity of P_{org} by the periphyton, the P_{org} transformation data were described using kinetic models (Pseudo-first-order and Pseudo-second-order kinetic equation). According to Fig. 2b, the amounts of P_{org} converted to P_{inorg} (q_e) after 48 h by the periphyton were 28.3, 23.9, and 22.5 mg g^{-1} under 0.2, 0.4, and 0.6 g L^{-1}, respectively. The pseudo-first-order and pseudo-second-order kinetic constants k and q values determined from the plots, decreased with the enhancement in the biomass of the periphyton (Table 1). This implies that the treatment with higher periphyton biomass contains a dense layer, resulting in smaller contribution on P_{org} transformation. This result is consistent with the observation in previous studies [27], [28]. It is notable that the fitting correlation coefficient (R^2) of pseudo-first-order model is better than that of the pseudo-second-order coefficient, suggesting that the pseudo-first-order kinetic model is more suitable for P_{org} transformation process.

To identify the role of phosphatase on P_{org} transformation by the periphyton, phosphatase activity under varied P_{org} concentration was evaluated. The phosphatase activity was increased initially but decreased subsequently as P_{org} concentration increased (Fig. 3). The maximal phosphatase activity was about 22 μmol pNP g^{-1} h^{-1} when the P_{org} concentration was 20 mg P L^{-1}. Phosphatase plays an important role in the biochemical cycles of phosphorus in aquatic system by hydrolyzing dissolved organic phosphorus to phosphates that are available for cellular uptake. Such enzymatic response to phosphorus limitation has been demonstrated previously in both planktonic communities and biofilms [21], [29], [30]. Our experiment also showed similar phosphatase activities, which demonstrates that phosphatase was a main factor affecting P_{org} conversion process.

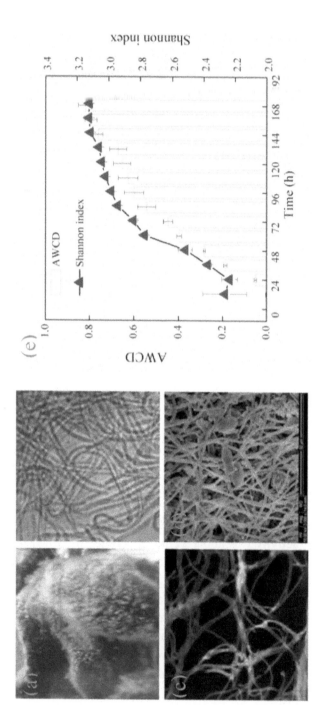

FIGURE 1: Characteristics of the the periphyton. The photo of the periphyton employed for the experiments (a), the periphyton observed under OM (b, ×2000), CLSM (c, ×2000), and SEM (d, ×2000); the microbial community diversities of the periphyton based on Biolog analyses (e).

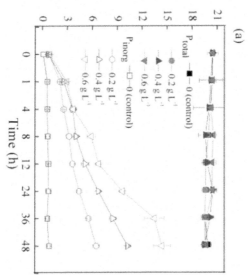

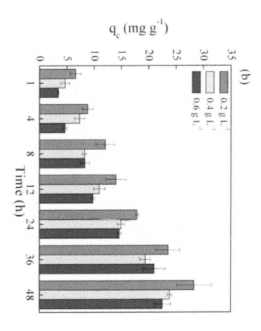

FIGURE 2: The conversion process of Porg (a) the change of the Ptotal and Pinorg over the time (b) the change of qc over the time (experiment conditions: light intensity = 12000 Lux, temperature = 25°C).

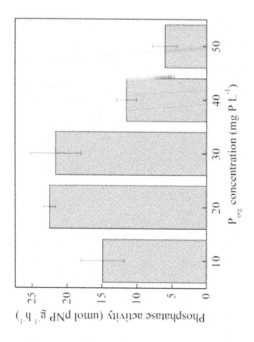

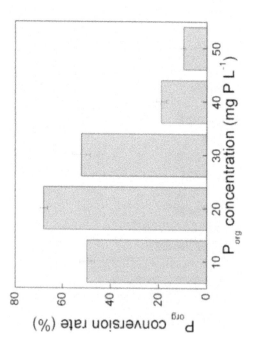

FIGURE 3: Phosphatase activity of the periphyton under different P_{org} concentrations.

TABLE 1: The kinetic parameters of Porg transformation by the periphyton.

Periphytom biofilm content	Pseudo-first-order kinetic model			Pseudo-two-order kinetic model		
	k_1 (h^{-1})	q_1 $(mg\ g^{-1})$	R^2	k_2 $(g\ mg^{-1}\ h^{-1})$	q_1 $(mg\ g^{-1})$	R^2
0.2 g L^{-1}	0.0432	24.632	0.964	0.0028	31.847	0.922
0.4 g L^{-1}	0.0416	21.463	0.977	0.0026	27.778	0.897
0.6 g L^{-1}	0.0412	20.888	0.984	0.0017	30.303	0.888

The phototrophic biofilm cultured under limited phosphate and organic phosphorus supply exhibited higher phosphatase activities than that cultured with sufficient phosphate supply [19], suggesting that end-product repression and de-repression of phosphatase activity was a main limitation factor of phosphatase activity. In this study, the P_{inorg} concentration transformed from P_{org} increased slowly when the periphyton mass was under 0.6 g L^{-1} in late time (after the 36 h) (Fig. 2).This result is similar with Ellwood's (2012) study. However, it was found that the P_{org} concentration also was an important factor determining the phosphatase activity of phototrophic periphyton (Fig. 3). There is a significant reduction in phosphatase activity under high P_{org} concentration ($p<0.05$). One possible explanation could be the growth repression under high P_{org} concentration, which showed that the periphyton were not functioning after being cultured in high P_{org} concentration (i.e., over 30 mg P L^{-1}).

The conversion from P_{org} to P_{inorg} is an important process for phosphorus removal and recovery since P_{inorg} is the removable and recoverable form of phosphorus in wastewater-treatment system. Many measures such as the advanced-oxidation processes (AOPs) have been applied for converting P_{org} to P_{inorg}, which are regarded as promising means for the transformation of P_{org} in the low-concentration streams [11]. AOPs rely on nonspecific free-radical species, such as hydroxyl radicals, to quickly attack the structure of organic compounds. However, this method mostly applied for the destruction of specific P-based and trace contaminants, such as organophosphorus pesticides [31]. Therefore, the AOPs methods might be impractical to apply in surface waters such as stream and lake, which often have high organic pollution loadings such as animal discharge. Compare to AOPs methods, many advantages such as low capital cost, easy-harvest, and powerfully converting ability for high content and non-specific P_{org} (ATP), suggesting that periphyton-based conversion system is a potential promising technology for P_{org} removal and recovery.

3.3.3 P_{ORG} REMOVAL BY THE PERIPHYTON

The P_{total} concentration in solution was decreased over time from 13 mg P L^{-1} to 6, 2, 0, and 0 mg P L^{-1} respectively under the treatments with the periphyton masses of 0.4, 0.8, 1.6, and 2.4 g L^{-1} respectively while the P_{total} content in the control showed slight reduction (Fig. 4). This indicates that the periphyton could remove P_{org} from artificial non-point source wastewater effectively. Simultaneously, the P_{inorg} concentration in solution showed the same change under varied biomass content treatments, which was firstly increased and then decreased. This indicates that the conversion of P_{org} by the periphyton occurred throughout the whole removal process. However, according to the P_{org} conversion trial (Fig. 2a), the concentrations of P_{inorg} transformed from P_{org} in solution should increase continuously over times and are larger than the P_{inorg} contents determined in Fig. 4. This implies that removing P_{inorg} that converted from P_{org} is a key procedure in the P_{org} removal process by the periphytons.

According to Fig. 5a, the removal rate under the four periphyton biomass levels (0.4, 0.8, 1.6, and 2.4 g L^{-1}) demonstrated a consistent trend over time within 48 h, and increased from 5%, 13%, 20%, and 29% to 54%, 80%, 100%, and 100%, respectively. This implies that the higher the content of the periphyton is available for P_{org} removal, the greater the amount of P_{org} is removed. Furthermore, to determine whether assimilation mechanism dominated the removal process of P_{org} by the periphyton, NaN_3 was used to impede microbial activity of the periphyton by restraining microbial respiration and inhibiting assimilation [32], [33]. The removal rates of P_{org} by the periphyton under NaN_3 treatment within 48 h were not significantly different from the controls (p>0.05, Fig. 5b), which indicates that the assimilation of phosphorus by microbes was minimal during the removing of P_{org} by the periphyton in 48 h. This further suggests that the P_{org} removal process of the periphyton is dominated by adsorption within 48 h.

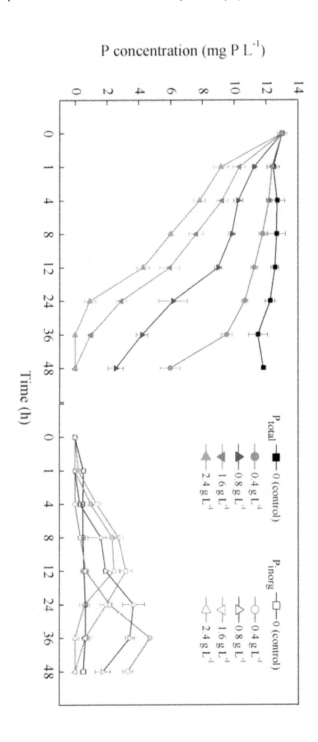

FIGURE 4: The removal process of P $_{org}$ by the periphyton. P $_{total}$ means total phosphorus content and P $_{inorg}$ means inorganic phosphorus content.

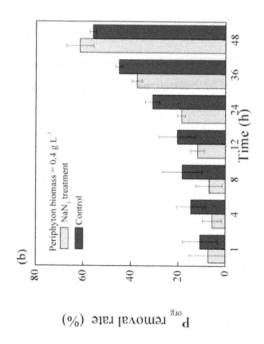

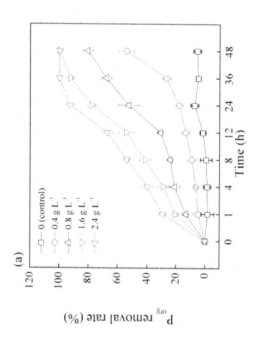

FIGURE 5: The P_{org} removal rate under different treatments.

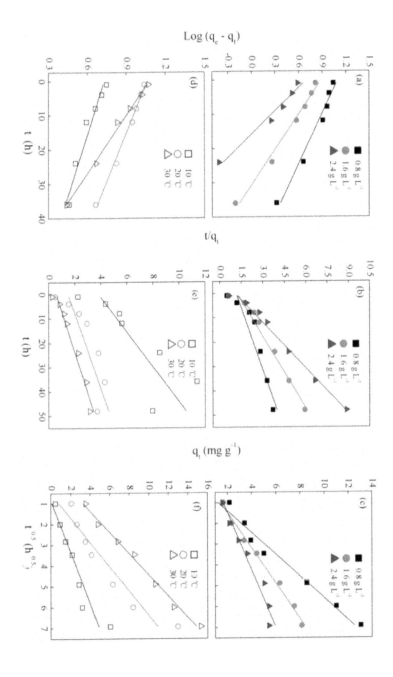

FIGURE 6: Adsorption kinetic analysis, (a d) the pseudo first-order kinetic and (b e) the pseudo second-order kinetic and (c f) the intra-particle diffusion kinetic of the periphyton biofilm for the P org with different biomass content at different temperature.

For a pseudo-second-order model, the correlation coefficient (R^2) is generally less than the pseudo-first-order coefficient (Table 2). Accordingly, kinetic parameters k_1 and q_1 showed the same trend that increased from 0.019 and 5.84 to 0.041and 11.97 respectively with temperature rise, while k_1 increased from 0.047 to 0.102 and q_1 decreased from 12.57 to 4.99 with the periphyton content increased. In view of these results, it can be safely concluded that the pseudo-first-order kinetic model provides a better correlation and description for the adsorption process of P_{org} by the periphyton at different temperatures and biomass (Fig. 6).

TABLE 2: Kinetic parameters for P_{org} removal by the periphyton.

Treat-ments	Pseudo-first-order kinetic			Pseudo-two-order kinetic			Intra-particle diffusion		
	k_1 (h^1)	q_1 (mg g^1)	R^2	k_2 (g mg^1 h^1)	q_1 (mg g^1)	R^2	k_{id} (mg g^1 h$^{0.5}$)	C	R^2
10°C	0.019	5.84	0.944	0.005	7.15	0.670	0.810	−0.71	0.876
20°C	0.025	11.51	0.911	0.003	15.34	0.707	1.707	−0.91	0.901
30°C	0.041	11.97	0.994	0.006	16.95	0.953	1.955	1.31	0.991
0.8 g L^{-1}	0.047	12.57	0.972	0.003	16.86	0.843	1.908	−0.66	0.973
1.6 g L^{-1}	0.067	7.74	0.986	0.009	9.85	0.961	1.173	0.27	0.990
2.4 g L^{-1}	0.102	4.99	0.973	0.024	6.22	0.966	0.719	0.95	0.953

In a solid-liquid system, most adsorption reactions take place through multi-step mechanisms, which at least contain external film diffusion, intra-particle diffusion, and interaction between adsorbate and active site. Thus, an intra-particle diffusion model was chosen to analyze the process of P_{org} adsorption onto the periphyton. The determination coefficients (R^2) were increased from 0.88 to 0.99 as temperature rose (Table 2), which suggested intra-particle diffusion may be rate controlled step under high temperature (30°C). The relatively high R^2 under different biomass contents (Table 2) implies that intra-particle diffusion in adsorption process of P_{org} by the periphyton was influenced by biomass. According to intra-particle diffusion model, if the plot of q_t versus $t^{0.5}$ presents a multi-linearity correlation, it indicates that three steps occur during the adsorption process: the first is the transport of molecules from the bulk solution to the adsorbent external surface by diffusion through the boundary layer (film diffusion). The second portion is the diffusion of the molecules from the external surface into the pores of the adsorbent. The third portion is the final equilibrium stage, where the molecules are adsorbed on the active sites on the internal surface of the pores and the intra-particle diffusion starts to slow down due to the solute concentration becoming lower [34], [35]. It was shown that the plot of qt versus $t^{0.5}$ presents a multi-linearity correlation and does not pass through the origin under low temperature (Fig. 6f), which indicates the adsorption of P_{org} by the periphyton was control by some other processes than intra-particle diffusion process under relatively low temperature. The large intercept (C) suggests that the process is largely of surface adsorption. This implies that the adsorption of P_{org} by the periphyton at temperature of 30°C and biomass of 2.4 g L^{-1} were more inclined to surface adsorption (Table 2).

To further reveal the types of P_{org} adsorption (physical and chemical) by the periphyton, Arrhenius equation was chosen to calculate the activation energy (E_a) based on kinetic parameters. The magnitude of E_a may give an idea about the type of adsorption. Two main types of adsorption may occur, physical and chemical. In physical adsorption, E_a value is usually low between 5–40 kJ mol^{-1} since the equilibrium is usually rapidly attained and the energy requirements are weak [36]. Chemical adsorption is specific and involves forces much stronger than physical adsorption, where E_a value is commonly high between 40 and 800 kJ mol^{-1} according to Arrhenius equation [37]. However, in some systems the chemical adsorption occurs very rapidly and E_a was relatively low, which is termed as a non-activated chemisorption [38].

The correlation coefficient of corresponding linear plot of ln k against 1/T is 0.96. The E_a value for the adsorption of P_{org} onto the periphyton is found to be as 27.082 kJ mol^{-1}, which suggests that the adsorption of P_{org} in the presence of the periphyton is exhibited the characteristic of physical adsorption.

Organic phosphorus can be found commonly in municipal, agricultural, and animal wastewaters, but there is scant information on its removal and recovery due to current phosphorus removal techniques are typically for inorganic phosphorus. Furthermore, as phosphorus resources becomes more scarce recently, phosphorus recovery from wastewaters by algal and macrophyte are regarded as a promising strategy and already in widespread use [39]. Compared to algal and macrophyte, the periphyton are more easy to be acquired and harvested. Therefore, the development of removing and capturing phosphorus from non-point source wastewaters for reuse based on the periphyton are primarily important to agriculture in the near future. In this study, our experimental results reveal that the periphyton not only possesses substantial capacity in effective organic phosphorus removal, but also the great ability in converting organic phosphorus to inorganic phosphorus that are readily captured. Finally, there are many advantages of the periphyton itself—it is environmentally friendly, economically viable and operationally simple. Given the above advantages, this phosphorus removal, recovery and reusing technologies based on the periphyton will have vast practical potentials, although it is also dependent on numerous factors such as light, temperature, water column phosphorus concentration, water flow velocity, the growth stage and thickness of periphyton [40], [41]. Most importantly, the native conditions of wastewaters in natural system (especially in agricultural wastewaters) are more complicated and the adsorption process may be reversible by the periphyton under high flow conditions. In such conditions, whether the adsorbed phosphorus will be released into aquatic ecosystem later from the periphyton needs further investigation.

3.4 CONCLUSIONS

Phototrophic periphyton can produce large amounts of phosphatase that can facilitate the transformation of organic phosphorus to inorganic phosphorus. This conversion process is influenced by the concentration of organic phosphorus and periphyton biomass in solution. Also, the periphyton can effectively remove organic phosphorus from artificial non-point source wastewaters, and the removal process is dominated by biosorption that exhibits the characteristic of physical adsorption. This bioadsorption process is distinct from the non-bioadsorption, which is more influenced by the environmental conditions such as temperature. This work gives an insight into the organic phosphorus conversion and removal processes in the presence of the periphyton or other similar microbial aggregates, contribute to the full understanding of phosphorus biogeochemical circulation in aquatic system, and provide kinetic data for the design of engineering of phosphorus removal, recovery and reusing technologies based on the periphyton.

REFERENCES

1. Burkholder JM, Noga EJ, Hobbs CH, Glasgow HB (1992) New 'phantom' dinoflagellate is the causative agent of major estuarine fish kills. Nature 358: 407–410. doi: 10.1038/358407a0
2. Wu Y, Kerr PG, Hu Z, Yang L (2010) Eco-restoration: Simultaneous nutrient removal from soil and water in a complex residential–cropland area. Environmental Pollution 158: 2472–2477. doi: 10.1016/j.envpol.2010.03.020
3. Wu Y, Li T, Yang L (2012) Mechanisms of removing pollutants from aqueous solutions by microorganisms and their aggregates: A review. Bioresource Technology 107: 10–18. doi: 10.1016/j.biortech.2011.12.088
4. McKelvie ID (2005) Separation, preconcentration and speciation of organic phosphorus in environmental samples; Turner BL, Frossard E, Baldwin DS, editors. London: CAB International.
5. Herbes SE, Allen HE, Mancy KH (1975) Enzymatic Characterization of Soluble Organic Phosphorus in Lake Water. Science 187: 432–434. doi: 10.1126/science.187.4175.432
6. Minear RA (1972) Characterization of naturally occurring dissolved organophosphorus compounds. Environmental Science & Technology 6: 431–437. doi: 10.1021/es60064a007
7. Turner BL, Cade-Menun BJ, Condron LM, Newman S (2005) Extraction of soil organic phosphorus. Talanta 66: 294–306. doi: 10.1016/j.talanta.2004.11.012
8. Cotner Jr JB, Wetzel RG (1992) Uptake of dissolved inorganic and organic phosphorus compounds by phytoplankton and bacterioplankton. Limnology and Oceanography 37: 232–243. doi: 10.4319/lo.1992.37.2.0232
9. Dyhrman ST, Chappell PD, Haley ST, Moffett JW, Orchard ED, et al. (2006) Phosphonate utilization by the globally important marine diazotroph Trichodesmium. Nature 439: 68–71. doi: 10.1038/nature04203
10 Sanudo-Wilhelmy SA (2006) Oceanography: A phosphate alternative. Nature 439: 25–26. doi: 10.1038/439025a
11. Rittmann BE, Mayer B, Westerhoff P, Edwards M (2011) Capturing the lost phosphorus. Chemosphere 84: 846–853. doi: 10.1016/j.chemosphere.2011.02.001
12. Donlan RM (2002) Biofilms: microbial life on surfaces. Emerg Infect Dis 8: 881–890. doi: 10.3201/eid0809.020063
13. Battin TJ, Kaplan LA, Denis Newbold J, Hansen CME (2003) Contributions of microbial biofilms to ecosystem processes in stream mesocosms. Nature 426: 439–442. doi: 10.1038/nature02152

14. Sabater S, Guasch H, Roman A, Muñoz I (2002) The effect of biological factors on the efficiency of river biofilms in improving water quality. Hydrobiologia 469: 149–156. doi: 10.1023/a:1015549404082

15. McCormick PV, Shuford RBE, Chimney MI (2006) Periphyton as a potential phosphorus sink in the Everglades Nutrient Removal Project. Ecological Engineering 27: 279–289. doi: 10.1016/j.ecoleng.2006.05.018

16. Guzzon A, Bohn A, Diociaiuti M, Albertano P (2008) Cultured phototrophic biofilms for phosphorus removal in wastewater treatment. Water Research 42: 4357–4367. doi: 10.1016/j.watres.2008.07.029

17. Scinto LJ, Reddy KR (2003) Biotic and abiotic uptake of phosphorus by periphyton in a subtropical freshwater wetland. Aquatic Botany 77: 203–222. doi: 10.1016/s0304-3770(03)00106-2

18. Gatidou G, Iatrou E (2011) Investigation of photodegradation and hydrolysis of selected substituted urea and organophosphate pesticides in water. Environmental Science and Pollution Research 18: 949–957. doi: 10.1007/s11356-011-0452-1

19. Ellwood NTW, Di Pippo F, Albertano P (2012) Phosphatase activities of cultured phototrophic biofilms. Water Research 46: 378–386. doi: 10.1016/j.watres.2011.10.057

20. Balser TC, Wixon DL (2009) Investigating biological control over soil carbon temperature sensitivity. Global Change Biology 15: 2935–2949. doi: 10.1111/j.1365-2486.2009.01946.x

21. Bentzen E, Taylor W, Millard E (1992) The importance of dissolved organic phosphorus to phosphorus uptake by limnetic plankton. Limnology and Oceanography 37: 217–231. doi: 10.4319/lo.1992.37.2.0217

22. Hill MO (1973) Diversity and evenness: a unifying notation and its consequences. Ecology 54: 427–432. doi: 10.2307/1934352

23. Dawood S, Sen TK (2012) Removal of anionic dye Congo red from aqueous solution by raw pine and acid-treated pine cone powder as adsorbent: Equilibrium, thermodynamic, kinetics, mechanism and process design. Water Research 46: 1933–1946.

24. de Beer D, Stoodley P (2006) Microbial biofilms. Prokaryotes 1: 904–937. doi: 10.1007/0-387-30741-9_28

25. De Beer D, Stoodley P, Roe F, Lewandowski Z (1994) Effects of biofilm structures on oxygen distribution and mass transport. Biotechnology and bioengineering 43: 1131–1138. doi: 10.1002/bit.260431118

26. James GA, Beaudette L, Costerton JW (1995) Interspecies bacterial interactions in biofilms. Journal of Industrial Microbiology 15: 257–262. doi: 10.1007/bf01569978

27. de los Ríos A, Ascaso C, Wierzchos J, Fernández-Valiente E, Quesada A (2004) Microstructural Characterization of Cyanobacterial Mats from the McMurdo Ice Shelf, Antarctica. Appl Environ Microbiol 70: 569–580. doi: 10.1128/aem.70.1.569-580.2004

28. Roeselers G, Loosdrecht M, Muyzer G (2008) Phototrophic biofilms and their potential applications. Journal of Applied Phycology 20: 227–235. doi: 10.1007/s10811-007-9223-2

29. Espeland EM, Wetzel RG (2001) Effects of photosynthesis on bacterial phosphatase production in biofilms. Microbial Ecology 42: 328–337. doi: 10.1007/s002480000117

30. Huang C-T, Xu KD, McFeters GA, Stewart PS (1998) Spatial patterns of alkaline phosphatase expression within bacterial colonies and biofilms in response to phosphate starvation. Applied and environmental microbiology 64: 1526–1531.

31. Badawy MI, Ghaly MY, Gad-Allah TA (2006) Advanced oxidation processes for the removal of organophosphorus pesticides from wastewater. Desalination 194: 166–175. doi: 10.1016/j.desal.2005.09.027

32. Saisho D, Nakazono M, Tsutsumi N, Hirai A (2001) ATP synthesis inhibitors as well as respiratory inhibitors increase steady-state level of alternative oxidase mRNA in Arabidopsis thaliana. Journal of Plant Physiology 158: 241–245. doi: 10.1078/0176-1617-00193

33. Wu Y, He J, Yang L (2010) Evaluating adsorption and biodegradation mechanisms during the removal of microcystin-rr by periphyton. Environmental Science & Technology 44: 6319–6324. doi: 10.1021/es903761y

34. Hameed BH, El-Khaiary MI (2008) Kinetics and equilibrium studies of malachite green adsorption on rice straw-derived char. Journal of Hazardous Materials 153: 701–708. doi: 10.1016/j.jhazmat.2007.09.019

35. Sun Q, Yang L (2003) The adsorption of basic dyes from aqueous solution on modified peat–resin particle. Water Research 37: 1535–1544. doi: 10.1016/s0043-1354(02)00520-1

36. Aksakal O, Ucun H (2010) Equilibrium, kinetic and thermodynamic studies of the biosorption of textile dye (Reactive Red 195) onto Pinus sylvestris L. Journal of Hazardous Materials 181: 666–672. doi: 10.1016/j.jhazmat.2010.05.064

37. Doğan M, Alkan M, Demirbaş Ö, Özdemir Y, Özmetin C (2006) Adsorption kinetics of maxilon blue GRL onto sepiolite from aqueous solutions. Chemical Engineering Journal 124: 89–101. doi: 10.1016/j.cej.2006.08.016

38. Kay M, Darling GR, Holloway S, White JA, Bird DM (1995) Steering effects in non-activated adsorption. Chemical Physics Letters 245: 311–318. doi: 10.1016/0009-2614(95)00975-a

39. Shilton AN, Powell N, Guieysse B (2012) Plant based phosphorus recovery from wastewater via algae and macrophytes. Current Opinion in Biotechnology 23: 884–889. doi: 10.1016/j.copbio.2012.07.002

40. Matheson FE, Quinn JM, Martin ML (2012) Effects of irradiance on diel and seasonal patterns of nutrient uptake by stream periphyton. Freshwater Biology 57: 1617–1630. doi: 10.1111/j.1365-2427.2012.02822.x

41. McCormick PV, O'Dell MB, Shuford Iii RBE, Backus JG, Kennedy WC (2001) Periphyton responses to experimental phosphorus enrichment in a subtropical wetland. Aquatic Botany 71: 119–139. doi: 10.1016/s0304-3770(01)00175-9

PART II

FOOD AND BEVERAGE INDUSTRIES

CHAPTER 4

Treatment of Slaughterhouse Wastewater in a Sequencing Batch Reactor: Performance Evaluation and Biodegradation Kinetics

PRADYUT KUNDU, ANUPAM DEBSARKAR, AND SOMNATH MUKHERJEE

4.1 INTRODUCTION

The continuous drive to increase meat production for the protein needs of the ever increasing world population has some pollution problems attached. Pollution arises from activities in meat production as a result of failure in adhering to Good Manufacturing Practices (GMP) and Good Hygiene Practices (GHP) [1]. Consideration is hardly given to safety practices during animal transport to the abattoir, during slaughter and dressing of hides and flesh [2]. For hygienic reasons abattoirs, use large amount of water in processing operations (slaughtering and cleaning), which produces large amount of wastewater. The major environmental problem associated with this abattoir wastewater is the large amount of suspended solids and

Treatment of Slaughter House Wastewater in a Sequencing Batch Reactor: Performance Evaluation and Biodegradation Kinetics. © *Kundu P, Debsarkar A, and Mukherjee S.* BioMed Research International **2013** *(2013), http://dx.doi.org/10.1155/2013/134872. Licensed under a Creative Commons Attribution 3.0 Unported License, http://creativecommons.org/licenses/by/3.0/.*

liquid waste as well as odor generation [3]. Effluent from slaughterhouses has also been recognized to contaminate both surface and groundwater because during abattoir processing, blood, fat, manure, urine, and meat tissues are lost to the wastewater streams [4]. Leaching into groundwater is a major part of the concern, especially due to the recalcitrant nature of some contaminants [5]. Blood, one of the major dissolved pollutants in abattoir wastewater, has the highest COD of any effluent from abattoir operations. If the blood from a single cow carcass is allowed to discharge directly into a sewer line, the effluent load would be equivalent to the total sewage produced by 50 people on average day [6]. The major characteristics of abattoir wastes are high organic strength, sufficient organic biological nutrients, adequate alkalinity, relatively high temperature (20 to 30°C) and free of toxic material. Abattoir wastewaters with the previous characteristics are well suited to anaerobic treatment and the efficiency in reducing the BOD5 ranged between 60 and 90% [7]. The high concentration of nitrates in the abattoir wastewater also exhibits that the wastewater could be treated by biological processes. Nitrogenous wastewater when discharged to receiving water bodies leads to undesirable problems such as algal blooms and eutrophication in addition to oxygen deficit. The dissolved oxygen level further depleted if organic carbon along with nutrient sinks into the water environment. Hence, it is very much necessary to control the discharge of combined organic carbon and nitrogen laden wastewater by means of appropriate treatment. Biological treatment has been proved to be comparatively innocuous and more energy efficient of treating wastewater if good process control could be ensured [8]. Several researchers successfully used different technologies for treatment of slaughterhouse wastewater containing organic carbon and nitrogen (COD and TKN) in laboratory and pilot scale experiment. Table 1 had shown the previous research findings about slaughterhouse wastewater treatment by the different investigators.

Among the various biological treatment processes, sequencing batch reactor (SBR) is considered to be an improved version of activated sludge process, which operates in fill and draw mode for biological treatment of wastewater. An SBR operates in a pseudobatch mode with aeration and sludge settlement both occurring in the same tank. SBRs are operated in fill-react-settle-draw-idle period sequences. The major differences between SBR and conventional continuous-flow, activated sludge system is that the SBR tank carries out the functions of equalization, aeration, and sedimentation in a time sequence rather than in the conventional space sequence of continuous-flow systems. Sequencing batch reactors (SBRs) are advocated as one of the best available techniques (BATs) for slaughterhouse wastewater treatment [16, 17] because they are capable of removing organic carbon, nutrients, and suspended solids from wastewater in a single tank and also have low capital and operational costs.

TABLE 1: Different available technologies used to treat slaughterhouse wastewater.

Sl no.	Technology adopted	Input characteristics of slaughterhouse wastewater	Observations	References
(1)	Anaerobic treatment of slaughterhouse wastewaters in a UASB (Upflow Anaerobic Sludge Blanket) reactor and in an anaerobic filter (AF).	Slaughterhouse wastewater showed the highest organic content with an average COD of 8000 mg/L, of which 70% was proteins. The suspended solids content represented between 15 and 30% of the COD.	The UASB reactor was run at OLR (Organic Loading Rates) of 1–6.5 kg COD/m³/day. The COD removal was 90% for OLR up to 5 kg COD/m³/day and 60% for an OLR of 6.5 kg COD/m³/day. For similar organic loading rates, the AF showed lower removal efficiencies and lower percentages of methanization.	Ruiz et al. [9]
(2)	Anaerobic sequencing batch reactors.	Influent total chemical oxygen demand (TCOD) ranged from 6908 to 11 500 mg/L, of which approximately 50% were in the form of suspended solids (SS).	Total COD was reduced by 90% to 96% at organic loading rates (OLRs) ranging from 2.07 to 4.93 kg m⁻³ d⁻¹ and a hydraulic retention time of 2 days. Soluble COD was reduced by over 95% in most samples.	Massé and Masse [10]
(3)	Moving bed sequencing batch reactor for piggery wastewater treatment.	COD, BOD, and suspended solids in the range of 4700–5900 mg/L, 1500–2300 mg/L, and 4000–8000 mg/L, respectively.	COD and BOD removal efficiency was greater than 80% and 90%, respectively at high organic loads of 1.18–2.36 kg COD/m³ d. The moving-bed SBR gave TKN removal efficiency of 86–93%.	Sombatsompop et al. [11]

TABLE 1: *Cont.*

Sl no.	Technology adopted	Input characteristics of slaughterhouse wastewater	Observations	References
(4)	Fixed bed sequencing batch reactor (FBSBR).	The wastewater has COD loadings in the range of 0.5–1.5 Kg COD/m³ per day.	COD, TN, and phosphorus removal efficiencies were at range of 90–96%, 60–88%, and 76–90%, respectively.	Rahimi et al. [12]
(5)	Chemical coagulation and electrocoagulation techniques.	COD and BOD5 of raw wastewater in the range of 5817 ± 473 and 2543 ± 362 mg/L.	Removal of COD and BOD5 more than 99% was obtained by adding 100 mg/L PACl and applied voltage 40 V.	Bazrafshan et al. [13]
(6)	Hybrid upflow anaerobic sludge blanket (HUASB) reactor for treating poultry slaughterhouse wastewater.	Slaughterhouse wastewater showed total COD 3000–4800 mg/L, soluble COD 1030–3000 mg/L, BOD5 750–1890 mg/L, suspended solids 300–950 mg/L, alkalinity (as $CaCO_3$) 600–1340 mg/L, VFA (as acetate) 250–540 mg/L, and pH 7–7.6.	The HUSB reactor was run at OLD of 19 kg COD/m³/day and achieved TCOD and SCOD removal efficiencies of 70–86% and 80–92%, respectively. The biogas was varied between 1.1 and 5.2 m³/m³ d with the maximum methane content of 72%.	Rajakumar et al. [14]
(7)	Anaerobic hybrid reactor was packed with light weight floating media.	COD, BOD and Suspended Solids in the range of 22000–27500 mg/L, 10800–14600 mg/L, and 1280–1500 mg/L, respectively.	COD and BOD reduction was found in the range of 86.0–93.58% and 88.9–95.71%, respectively.	Sunder and Satyanarayan [15]

Biological treatment of wastewater containing organic carbon and nitrogen (COD and TKN) is also carried out in laboratory and pilot scale experiment by several researchers successfully [18–26]. Nutrients in piggery wastewater with high organic matter, nitrogen, and phosphorous content were biological removed by Obaja et al. [27] in a sequencing batch reactor (SBR) with anaerobic, aerobic, and anoxic stages. The SBR was operated with wastewater containing 1500 mg/L ammonium and 144 mg/L phosphate, a removal efficiency of 99.7% for nitrogen and 97.3% for phosphate was obtained. A full-scale SBR system was evaluated by Lo and Liao [28] to remove 82% of BOD and more than 75% of nitrogen after a cycle period of 4.6 hour from swine wastewater. Mahvi et al. [29] carried out a pilot-scale study on removal of nitrogen both from synthetic and domestic wastewater in a continuous flow SBR and obtained a total nitrogen and TKN removal of 70–80% and 85–95%, respectively. An SBR system demonstrated by Lemaire et al. [30] to high degree of biological remove of nitrogen, phosphorus, and COD to very low levels from slaughterhouse wastewater. A high degree removal of total phosphorus (98%), total nitrogen (97%), and total COD (95%) was achieved after a 6-hour cycle period. Moreover, SBRs have been successfully used to treat landfill leachate, tannery wastewater, phenolic wastewater, and various other industrial wastewaters [31–34].

In the present investigation, an attempt has been made to explore the performance efficacy of SBR technology for simultaneous removal of soluble carbonaceous organic matter and ammonia nitrogen from slaughterhouse wastewater and also to determine the biokinetic constants for carbon oxidation, nitrification, and denitrification under different combination of react periods (aerobic/anoxic).

4.2 MATERIAL AND METHODS

4.2.1 SEED ACCLIMATIZATION FOR COMBINED CARBON OXIDATION AND NITRIFICATION

The active microbial seed was cultured under ambient condition in the laboratory by inoculating 200 mL sludge as collected from an aeration pond of M/S Mokami small-scale slaughterhouse located in the village Nazira, South 24 Parganas district (West Bengal), India, to a growth propagating media composed of 500 mL dextrose solution having concentrations of 1000 mg/L, 250 mL ammonium chloride (NH_4Cl) solution having concentration of 200 mg/L and 250 mL of nutrient solution in 3000 mL capacity cylindrical vessel. The composition of the nutrient solution in 250 mL distilled water was comprised of 60.0 mg K_2HPO_4, 40.0 mg KH_2PO_4, 500.0 mg $MgSO_4 \cdot 7H_2O$, 710.0 mg $FeCl_3 \cdot 6H_2O$, 0.1 mg $ZnSO4 \cdot 7H2O$, 0.1 mg $CuSO_4 \cdot 5H_2O$, 8.0 mg $MnCl_2 \cdot 2H_2O$, 0.11 mg $(NH_4)6Mo_7O_{24}$, 100.0 mg $CaCl_2 \cdot 2H_2O$, 200.0 mg $CoCl_2 \cdot 6H_2O$, 55.0 mg $Al_2(SO_4)_3 \cdot 16H_2O$,

150.0 mg H_3BO_3. Finally 800 mL volume of distilled water was added to liquid mixture to make a volume of 2 L and the mixture was continuously aerated with intermittent feeding with dextrose solution having concentrations of 1000 mg/L and ammonium chloride (NH_4Cl) having concentration of 200 mg/L as a carbon and nitrogen source, respectively. The acclimatization process was continued for an overall period of 90 days. The biomass growth was monitored by the magnitude of sludge volume index (SVI) and mixed liquor suspended solid (MLVSS) concentration in the reactor. pH in the reactor was maintained in the range 6.8–7.5 by adding required amount of sodium carbonate (Na_2CO_3) and phosphate buffer. The seed acclimatization phase was considered to be over when a steady-state condition was observed in terms of equilibrium COD and NH_4^+-N reduction with respect to a steady level of MLVSS concentration and SVI in the reactor.

Denitrifying seed was cultured separately in 2.0 L capacity aspirator bottle under anoxic condition. 500 gm of digested sludge obtained from the digester of a nearby sewage treatment plant (STP) was added to 1.0 L of distilled water. The solution was filtered and 250 mL of nutrient solution along with 250 mL dextrose solution as carbon source and 100 mL potassium nitrate solution (KNO_3) as the source of nitrate nitrogen (NO_3^--N) was added to it. The resulting solution was acclimatized for denitrification purpose under anoxic condition. Magnetic stirrer was provided for proper mixing of the solution. Denitrifying seed was acclimatized against a nitrate-nitrogen concentration varying from 10–90 mg/L as N, over a period of three months.

4.2.2 EXPERIMENTAL PROCEDURE

The experimental work was carried out in a laboratory scale SBR, made of Perspex sheet of 6 mm thickness, having 20.0 L of effective volume. In order to assess the treatability of slaughterhouse wastewater in an SBR, the real-life wastewater samples were collected from two different locations (i) the raw (untreated) wastewater from the main collection pit and (ii) the primary treated effluent from the inlet box of aeration basin. The wastewater samples were collected 4 (four) times over the entire course of the study in 25.0 L plastic containers and stored in a refrigerator at approximately 4.0°C. The effluent quality was examined as per the methods described in "Standard Methods" [35] for determining its initial characteristics which are exhibited in Table 2.

TABLE 2: Characteristics and composition of slaughterhouse wastewater.

Param- eters	Raw wastewater					Pretreated wastewater				
	Set-1	Set-2	Set-3	Set-4	Range	Set-1	Set-2	Set-3	Set-4	Range
pH	8.0	8.2	8.5	8.4	8.0–8.5	7.5	7.2	8.5	7.8	7.5–8.5
TSS (mg/L)	10120	12565	14225	13355	10120– 14225	2055	2280	2540	2386	2055– 2540
TDS (mg/L)	6345	7056	7840	6865	6345– 7840	2800	3065	3230	3185	2800– 3230
DO (mg/L)	0.8	1.1	0.9	1.3	0.8–1.3	1.2	1.4	1.5	1.6	1.2–1.6
SCOD (mg/L)	6185	6525	6840	6455	6185– 6840	830	945	1045	925	830– 1045
BOD5 at 20°C (mg/L)	3000	3200	3500	3350	3000– 3500	210	240	265	252	210– 265
TKN (mg/L as N)	1050	1130	1200	1165	1050– 1200	305	420	525	485	305– 525
$NH_4^+ N$ (mg/L as N)	650	695	735	710	650– 735	95	155	191	125	95–191

The settled effluent was poured in the reactor of 20.0 L capacity to perform necessary experiments. 2.5 L of preacclimatized mixed seed containing carbonaceous bacteria, nitrifier, and denitrifier was added in the reactor containing 20.0 L of pretreated slaughterhouse wastewater to carry out the necessary experiments. Oxygen was supplied through belt-driven small air compressor. A stirrer of 0.3 KW capacity was installed at the center of the vessel for mixing the content of the reactor. Air was supplied to the reactor during aerobic phase of react period with the help of diffused aeration system. However, during the anoxic phase the stirrer was allowed only to operate for mixing purpose and air supply was cut off. A timer was also connected to compressor for controlling the sequence of different react period (aerobic and anoxic). A schematic diagram of the experimental setup is shown in Figure 1.

The cycle period for the operation of SBR was taken as 10 hour, with a fill period of 0.5 hour, overall react period of 8.0 hours, settle period of 1.0 hour, and idle/decant period of 0.5 hour. The overall react period was divided into aerobic and anoxic react period in the following sequences: Combination-1: 4-hour aerobic react period and 4-hour anoxic react period. Combination-2: 5-hour aerobic react period and 3-hour anoxic react period. Combination-3: 3-hour aerobic react period and 5-hour anoxic react period. The performance study was carried out with pretreated slaughterhouse wastewater with same initial soluble chemical oxygen demand (SCOD) and two different ammonia nitrogen (NH_4^+-N) concentration of 100 ± 50 mg/L and 90 ± 10 mg/L, 1000 ± 50 mg/L and 180 ± 10 mg/L, respectively. During the fill period of 30 min duration, 16.0 L of slaughterhouse wastewater was transferred under gravity from a feeding tank into the reactor. The mechanical mixer was operated continuously with a speed of 400 rpm from the beginning of the fill phase till the end of the total react phase for proper mixing of liquid in the reactor. During the draw phase, the supernatant wastewater was decanted until the liquid volume in the reactor was decreased to 4.0 L. sludge retention time (SRT) was manually controlled by withdrawal of volume of the mixed liquor from the reactor every day at the onset of the commencement of settle phase. The reactor was continuously run for 120 days. The initial pH values in the reactor were kept in between 7.1 and 7.7, whereas the sludge volume index (SVI) has been kept within the range of 75–85 mL/gm, for obtaining good settling property of the biomass. It has been reported that SRT should be longer than 10 days to achieve efficient nitrogen removal [36]. The SRT of 20–25 days as maintained for carbon oxidation and nitrification in the present SBR system for treatment of wastewater as suggested by Tremblay et al. [37].

During the time course of the study, 100 mL of sample was collected from the outlet of the reactor at every 1.0 hour interval, till completion of the fill period. The samples were analyzed for the following parameters: pH, DO, MLSS, MLVSS, COD, NH_4^+-N, NO_2^--N, and NO_3^--N as per the methods described in "Standard Methods" [35]. The pH of the solution was measured by a digital pH meter. NH_4^+-N, NO_2^--N, and NO_3^--N were estimated by respective ion selective electrodes in ISE meter. COD was analyzed by closed reflux method using dichromate digestion principle in digester. Dissolved oxygen (DO) was measured electrometrically by digital DO meter. Mixed liquor suspended solids (MLSS) and Mixed liquor volatile suspended solids (MLVSS) were measured by gravimetric method at temperature of 103–105°C and 550 ± 50°C in muffle furnace, respectively.

4.2.3 CARBON OXIDATION AND NITRIFICATION KINETICS IN SBR

Biokinetic parameters play an important role in designing and optimizing an activated sludge process. The biokinetic constants describe the metabolic performance of the microorganisms when subjected to the substrate and other components of the specific wastewater. These biokinetic coefficients yield a set of realistic design parameters, which can be used in rationalizing the design of the activated sludge process for a specific substrate.

4.2.3.1 SUBSTRATE REMOVAL KINETICS

The substrate removal constants, namely, half saturation concentration (K_s) and the maximum rate of substrate utilization (k) were determined from the Lawrence and McCarty's modified Monod equation [38] given below:

$$\frac{1}{U} = \left(\frac{K_S}{k}\right)\left(\frac{1}{S}\right) + \frac{1}{k}$$

(1)

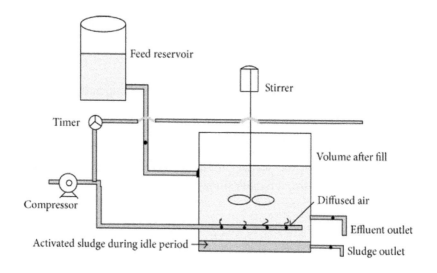

FIGURE 1: A schematic diagram of the experimental setup.

S = Substrate (SCOD and NH_4^+-N) concentration at any time in reactor (mg/L), U = Specific substrate utilization rate = $(S_0 - S)/\theta X$ (mg of SCOD or mg of NH_4^+-N/day/mg of MLVSS), θ = Contact time (day), X = MLVSS at any time in the reactor (mg/L), S_0 = Substrate (SCOD and NH_4^+-N) concentration of the influent (mg/L).

The plots made between 1/U and 1/S develops into a straight line with K_s/k as its slope and 1/k as its intercept.

4.2.3.2 SLUDGE GROWTH KINETICS

The sludge growth kinetic constants namely the yield coefficient (Y) and the endogenous decay coefficient (K_d), were determined from the Lawrence and McCarty's modified Monod equation [38] given below:

$$\frac{1}{\theta} = YU - K_d \tag{2}$$

where U = Specific substrate utilization rate (mg of SCOD or mg of NH_4^+-N/day/mg of MLVSS), θ = Contact time (day), k_d = Endogenous decay coefficient (day^{-1}), and Y = Yield coefficient (mg of MLVSS produced/mg of SCOD or NH_4^+-N).

A graph drawn between $1/\theta$ and U gives a straight line, with Y as its slope and k_d as its intercept.

4.2.4 DENITRIFICATION KINETICS IN SBR

In almost all cases denitrification started occurring at the onset of anoxic period and specific denitrification rate (q_{DN}) was calculated under different initial organic carbon and NH_4^+-N concentrations for different react period combinations, namely, (4+4), (5+3), (3+5) hrs over the respective anoxic environment.

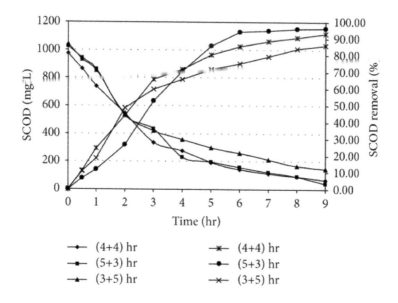

FIGURE 2: Carbon oxidation profile under different react period combination [Initial SCOD = 1000 ± 50 mg/L; Initial NH_4--N = 90 ± 10 mg/L as N].

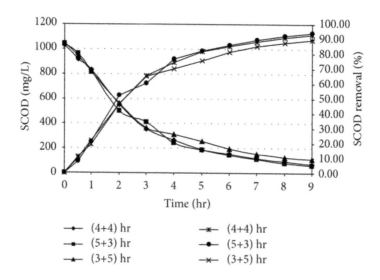

FIGURE 3: Carbon oxidation profile in SBR under different react period combination [Initial SCOD = 1000 ± 50 mg/L; Initial NH_4^+-N = 180 ± 10 mg/L as N].

4.3 RESULTS AND DISCUSSION

4.3.1 CARBON OXIDATION PERFORMANCE

Organic carbon, which is the source of energy for heterogenic and denitrifying microorganism, was estimated as chemical oxygen demand (COD). In the present experiment, in correspondance to an initial SCOD of 975.25 mg/L and initial NH_4^+-N concentration of 87.52 mg/L as N, it has been observed that the major fraction of SCOD removal took place within 4 or 5 hrs of aerobic react period. In anoxic phase, further SCOD removal has been noticed as shown in Figure 2. Li et al. [39] obtained that the maximum removal efficiency of COD (96%) for treatment of slaughterhouse wastewater which was marginally higher than the result of this present study. COD removal profile was also observed in similar pattern in the presence of higher initial NH_4^+-N concentration of 185.24 mg/L as N and initial SCOD of 1028.55 mg/L in a separate set of experiment. The results are plotted in Figure 3. It is revealed from Figures 2 and 3 that the rate of organics utilization by the dominant heterotrophs during initial aerobic react period was more as compared to its rate of removal during anoxic period. The carbon utilization bacteria used up bulk amount COD for energy requirement and growth. The removal efficiency of COD in the suspended growth reactor system depends on COD:TKN ratio. The mean COD:TKN ratio recommended for adequate carbon oxidation and nutrient removal as 10–12 [40]. In the present investigation, COD:TKN ratio was approximately 11.14 which was in agreement in their recommendation. The removal efficiency also depends on react time. The carbon utilizing bacteria obviously and is able to degrade more COD and produce CO_2 with production of new cells due to enhancement of aeration time. A marked improvement has been noticed for higher percentage removal of COD during increase of aeration time. A similar observation was noticed by Kanimozhi and Vasudevan [41]. Due to the increase of time and COD load more cells to be produced eventually higher degree of organic removal. When the react period was changed into 5-hour aerobic followed by a reduced 3.0 hour anoxic, a marginal improvement of SCOD removal in aerobic phase (77.27%) and anoxic phase (96.07%) with an initial SCOD of 1023.22 mg/L was observed due to enhanced aeration time. On the other hand, when the react period was subsequently changed to 3.0 hour, aerobic period followed by 5.0-hour anoxic period, a marginal decrease of SCOD removal in aerobic phase (65.64%) and anoxic phase (86.07%) with an initial SCOD of 1042.52 mg/L was obtained due to lag of aeration time.

4.3.2 NITRIFICATION PERFORMANCE

Ammonia oxidation took place due to the presence of previously acclimatized nitrifying organisms within the reactor as mixed culture. The nitrification results are shown in Figures 4 and 5. In case of specific cycle period of 4 hr (aerobic) and 4 hr (anoxic), it was observed that at the end of 8 hr react period of reaction, 90.12% nitrification could achieved for an initial NH_4^+-N was approximately 87.52 mg/L as Fongsatitkul et al. [40] obtained maximum 93% removal efficiency of soluble nitrogen for treatment of abattoir wastewater which was slightly higher than our result. The ammonia oxidation occurred in two phases; a fraction of ammonia was assimilated by cell-mass for synthesis of new cell during carbon oxidation and, in the subsequent phase, dissimilatory ammonia removal took place for converting NH_4^+-N into NO_2^--N and NO_3^--N under aerobic period. The dissimilatory removal of ammonia depends on the population of nitrifiers and oxidation time. The descending trend of ammonia removal for higher level of initial concentration of NH_4^+-N was attributed due to limitation of enzymatic metabolism of nitrifiers. When the reactor system was operated in 5 hr (aerobic) and 3 hr (anoxic) mode of react cycle, an overall performance of ammonia oxidation was improved from 90.12 to 96.20% and 84.41% for initial NH_4^+-N of 93.54 mg/L and 173.88 mg/L as N, respectively. Higher oxidation period was also recommended by earlier investigators [42, 43] for describing similar kind of experiment on landfill leachate treatment in SBR. The results reveal the fact that the extension of aeration period helped to enhance the oxidation efficiency for the present system. It was also observed that when aerobic period was reduced to 3.0 hr, ammonia oxidation reduced to 79.18% and 70.53% corresponding to initial NH_4^+-N value of 96.58 mg/L and 176.85 mg/L, respectively, at the end of 8 hr react period.

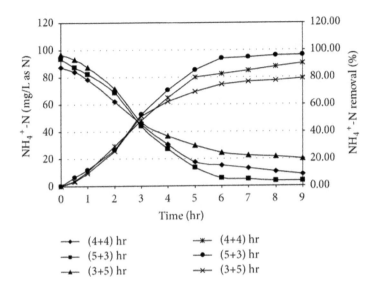

FIGURE 4: Ammonia oxidation profile in SBR under different react period combination [Initial SCOD = 1000 ± 50 mg/L; Initial NH$_4^+$-N = 90 ± 10 mg/L as N].

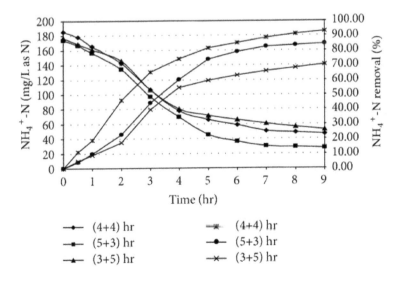

FIGURE 5: Ammonia oxidation profile in SBR under different react period combination [Initial SCOD = 1000 ± 50 mg/L; Initial NH$_4^+$-N = 180 ± 10 mg/L as N].

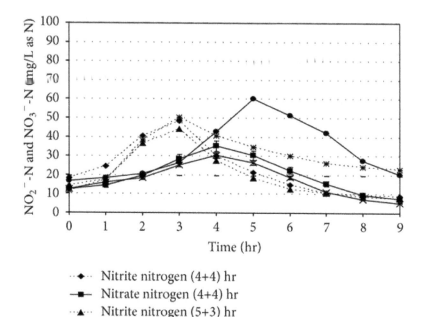

FIGURE 6: Nitrite and nitrate concentration profiles in SBR under different react period combination [initial SCOD = 1000 ± 50 mg/L and initial NH_4^+-N = 90 ± 10 mg/L as N].

4.3.3 DENITRIFICATION PERFORMANCE

The nitrite and nitrate nitrogen (NO_2^--N and NO_3^--N) level in the reactor during the total reaction period is shown in Figure 6. The maximum nitrite level was observed in between 2.5 and 3.0 hr of react period. The peak nitrate (NO_3^-) level was found to be formed close to 4.0 hr of aeration period for (4+4) and (3+5) hr combinations of react period. A time lag of one hour for maximum nitrate formation was also noticed after the

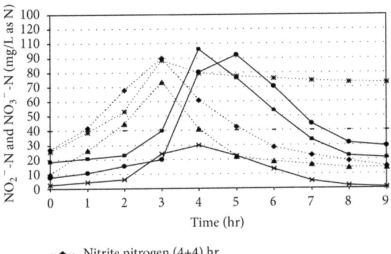

FIGURE 7: Nitrite and nitrate concentration profiles in SBR under different react period combination [initial SCOD = 1000 ± 50 mg/L and initial NH_4^+-N = 180 ± 10 mg/L as N].

attainment of the maximum NO_2^--N level in the reactor. For (5+3) hr react period combination, the formation of NO_3^- showed a time-dependent factor as the peak was found at the end of 5.0 hrs. In the Figure 6, after 4.0 hr of aeration period, the NO_3^- level was found to be 35.21 mg/L as N corresponding to initial NH_4^+-N level of 87.52 mg/L as N and NO_3^- concentration of 12.35 mg/L as N, respectively. On the other hand, after 5.0 hour of aerated react period, -N concentration in the reactor was found to be 60.24 mg/L as N for an initial NH_4^+-N and NO_3^--N concentration of 93.54

and 16.52 mg/L as N, respectively. The maximum NO_3^--N concentration for (3+5) hour react period combination was found to be 25.31 mg/L as N for the initial NH_4^+-N concentration of 96.58 mg/L as N and NO_3^--N level of 12.35 mg/L as N. The experimental results clearly indicate the necessity of longer aeration period for achieving maximum utilization of ammonia by the nitrifiers.

In Figure 7, after 4.0 hour of anoxic react period, nitrate (NO_3^-) was reduced to 22.29 mg/L as N from its peak concentration of 96.22 mg/L as N, which achieved a 76.83% removal of nitrate for initial NH_4^+-N concentration 185.24 mg/L as N. During denitrification phase, the residual soluble COD concentration as available was found to be more than the stoichiometric organic carbon requirement for effective denitrification. When the anoxic react period was reduced to 3.0 hr, it was observed that, nitrate concentration after 5 hr of aerobic period was found to be maximum (92.11 mg/L as N), per cent removal of nitrate descended from 76.83 to 66.16% for initial NH_4^+-N concentration 173.88 mg/L as N, due to insufficient of anoxic period.

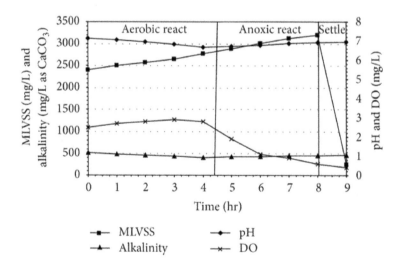

FIGURE 8: MLVSS, pH, alkalinity, and DO profiles for slaughterhouse wastewater treatment in SBR under (4+4) hr react period combination.

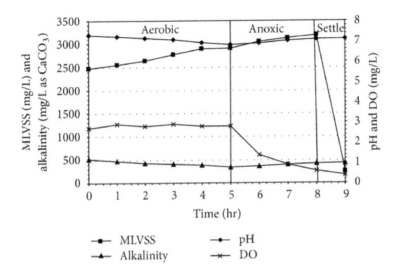

Figure 9: MLVSS, pH, alkalinity, and DO profiles for slaughterhouse wastewater treatment in SBR under (5+3) hr react period combination.

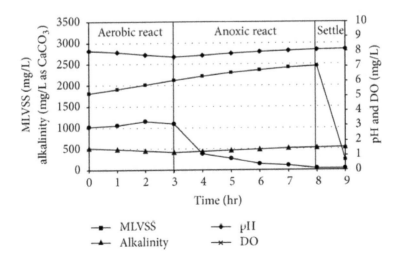

Figure 10: MLVSS, pH, alkalinity, and DO profiles for slaughterhouse wastewater treatment in SBR under (3+5) hr react period combination.

4.3.4 MLVSS, PH, ALKALINITY, AND DO PROFILES IN THE SBR DURING EXPERIMENT

The pH and alkalinity values of a biological system are vital parameters for microbial denitrification. The value of pH increases for ammonification and denitrification, decreases for organic carbon oxidation and nitrification. Alkalinity is not only important for nitrification and denitrification, but to also be used for indicating the system stability. Alkalinity was found to have a close correlation with SBR operating conditions, since different extents of nitrification (alkalinity consumption) and denitrification (alkalinity production) contribute to the variation of alkalinity in the system. During the aerobic phase, the minimal value of the pH curve was characterized the end of nitrification (Figures 8, 9, and 10). At the beginning of anoxic react phase, when ammonia nitrogen concentration was reduced considerably, pH starts to increase. This has occurred between 4.0 and 5.0 hr after the starting of aerobic react period in all experimental sets. The DO profile exhibited a sharp fall after which DO concentration decreased markedly at anoxic phase and reached minimum value. In the present, study the DO level remained almost steady during the entire aerobic react period with a marginal increase in DO level, but a marked descending trend was observed during the anoxic period in all the reaction sets irrespective of initial SCOD and ammonia concentrations. Under strict anaerobic condition the DO should be equal to zero, but anoxic environment starts from DO level less than 1.5 mg/L. At the start of anoxic react period most of the cases, DO was found to be less than 1.5 mg/L and at the end of anoxic react period the value becomes less than 1.0 mg/L.

4.3.5 KINETIC STUDY FOR ORGANIC CARBON REMOVAL FROM SLAUGHTERHOUSE WASTEWATER IN SBR

In the present study, the performance evaluation of the SBR system was also carried out from the view point of reaction kinetics determination for treating slaughterhouse wastewater. The values for the reciprocal of specific substrate utilization rate (1/U) were plotted against the reciprocal of effluent SCOD (1/S) and substrate removal kinetics was evaluated using

(1) as stated earlier. The slope of the straight line is (K_s/k) and intercept is $(1/k)$. The reciprocal of the contact time $(1/\theta)$ were plotted against the specific substrate utilization rate (U) and microbial growth kinetics was evaluated using (2). The yield coefficient (Y) was determined from the slope of the line and the endogenous decay coefficient (k_d) was obtained from intercept, $k_d = -C$. The values of biokinetic coefficients (k, K_s, Y, k_d) for combined carbon-oxidation and nitrification are listed in Table 3.

TABLE 3: Evaluation of biokinetic coefficients for carbon oxidation from slaughterhouse wastewater in SBR.

Initial SCOD (mg/L)	(4+4) hr react period combination	(5+3) hr react period combination	(3+5) hr react period combination	Standard values for kinetic constants [44]
1000 ± 50	(i) Substrate utilization—y= 70.32x + 0.215	(i) Substrate utilization—y = 68.22x + 0.187	(i) Substrate utilization—y = 42.65x + 0.285	K (day^{-1}) = (2–10)
	(ii) Microbial growth—y = 0.522x – 0.051	(ii) Microbial growth—y = 0.622x – 0.057	(ii) Microbial growth—y = 0.485x – 0.047	K_s (mg/L SCOD) = (15–70)
	Kinetic constants:	Kinetic constants:	Kinetic constants:	Y (mg VSS/ mg SCOD) = (0.4–0.8)
	k (day^{-1}) = 4.65	k (day^{-1}) = 5.34	k (day^{-1}) = 3.50	k_d (day^{-1}) = (0.025–0.075)
	K_s (mg/L SCOD) = 327.06	K_s (mg/L SCOD) = 364.81	K_s (mg/L SCOD) = 149.64	
	Y (mg VSS/mg SCOD) = 0.522	Y (mg VSS/mg SCOD) = 0.622	Y (mg VSS/mg SCOD) = 0.485	
	k_d (day^{-1}) = 0.051	k_d (day^{-1}) = 0.057	k_d (day^{-1}) = 0.047	

From Table 3, it has been estimated that the value of yield coefficient (Y) for the heterotrophs is in the range from 0.485 to 0.622. The yield coefficient was found to be improved with the increase in aeration period. The half velocity constant (K_s) values were found in the range of 149.64 to 364.81 for different combinations of react period. In the case of (5+3) combination of react cycle, the k and Y values are marginally higher than (4+4) and (3+5) combination. It was attributed to the fact that, after the initial acclimatization; the heterotrophs converted the carbon content at 5.0 hrs period of time more efficiently. After 5.0 hrs of aerobic react period, the available carbon content was reduced considerably and a fraction of heterotrophs attained endogenous state of condition while the nitrifiers are rejuvenated and started nitrification activity. This metabolism is also supported by the value of endogenous decay rate constant (k_d). In the case of (5|3) combination of react cycle k_d value is found to be 0.057 which is between 0.051 and 0.047 for the cases of (4+4) and (3+5) react period combinations, respectively. The values of biokinetic coefficients, other than K_s, such as k, y, k_d as obtained from the test results for carbon-oxidation and nitrification are also in congruence with their respective typical values [44].

4.3.6 KINETIC STUDY FOR AMMONIUM NITROGEN REMOVAL FROM SLAUGHTERHOUSE WASTEWATER IN SBR

The nitrification removal kinetics for mixed population (heterotrophs and nitrifiers) followed an identical pattern to organic carbon removal kinetics. A fraction of biological oxidation was attributed to the fact that a mixed population performed in the reactor along with nitrifiers. The linear graphs are plotted between (1/S) and (1/U) for substrate utilization kinetics under three different combinations of react period, namely, (4+4), (5+3), and (3+5), respectively, using (1). Microbial growth kinetics was evaluated using (2), which were determined by plotting straight lines between (1/θ) and (U) under three different combinations of react period, namely, (4+4), (5+3) and (3+5) hrs, respectively. The kinetic coefficient values for nitrification from the previous plots are given in Table 4. It has been clearly shown earlier that an increasing trend of higher removal efficiency for ammonia oxidation could be observed for extension of the aerobic react period beyond 4 hrs. This previous phenomenon also reflected the magnitudes of biokinetic constants under all experimental combinations of react period. The kinetic coefficients Y, k_d, and K_s were found to be in the range of 0.205 to 0.284, 0.037 to 0.051, and 21.83 to 70.93, respectively. The ammonia concentration found in the slaughterhouse wastewater was very high 180 ±10 mg/L as N for an inlet SCOD concentration of 1000 ± 50 mg/L, which are not usually present in any municipal wastewater stream. For this reason, the K_s value was found to be higher than the standard values (0.2–5.0 mg/L) considered for nitrification of municipal wastewater stream [44].

TABLE 4: Evaluation of biokinetic coefficients for nitrification from slaughterhouse wastewater in SBR.

Initial NH_4^+-N (mg/L as N)	(4+4) hr react period combination	(5+3) hr react period combination	(3+5) hr react period combination	Standard values for kinetic constants [44]
180 ± 10	(i) Substrate utilization—	(i) Substrate utilization—	(i) Substrate utilization—	$k \, (\text{day}^{-1}) = (1–30)$
	$y= 2.371x + 0.047$	$y = 2.412x + 0.034$	$y = 1.223x + 0.056$	$K_s \, (\text{mg/L } NH_4^+\text{-N}) = (0.2–5.0)$
	(ii) Microbial growth—	(ii) Microbial growth—	(ii) Microbial growth—	$Y \, (\text{mg VSS/ mg } NH_4^+\text{-N}) = (0.1–0.3)$
	$y = 0.234x - 0.047$	$y = 0.284x - 0.051$	$y = 0.205x - 0.037$	$k_d \, (\text{day}^{-1}) = (0.03–0.06)$
	$k \, (\text{day}^{-1}) = 21.27$	$k \, (\text{day}^{-1}) = 29.41$	$k \, (\text{day}^{-1}) = 17.85$	
	$K_s \, (\text{mg/L } NH_4^+\text{-N}) = 50.44$	$K_s \, (\text{mg/L } NH_4^+\text{-N}) = 70.93$	$K_s \, (\text{mg/L } NH_4^+\text{-N}) = 21.83$	
	$Y \, (\text{mg VSS/mg } NH_4^+\text{-N}) = 0.234$	$Y \, (\text{mg VSS/mg } NH_4^+\text{-N}) = 0.284$	$Y \, (\text{mg VSS/mg } NH_4^+\text{-N}) = 0.205$	
	$k_d \, (\text{day}^{-1}) = 0.047$	$k_d \, (\text{day}^{-1}) = 0.051$	$k_d \, (\text{day}^{-1}) = 0.037$	

4.3.7 DENITRIFICATION RATES FOR TREATMENT OF SLAUGHTERHOUSE WASTEWATER IN SBR

Specific denitrification rate (q_{DN}) was measured in terms of the rate of NO_3^--N removed per unit mass of denitrifying microorganisms, for three different react period combinations, namely, (4+4), (5+3), and (3+5) under the respective anoxic environment and the results are listed in Table 5. The specific denitrification rate (q_{DN}) is expressed on average basis spanningover respective anoxic periods of 3.0, 4.0, and 5.0 hours. The average specific denitrification rate (q_{DN}), in (5+3), (4+4), and (3+5) cases was found to increase considerably with the increase in average anoxic SCOD utilization rate (q_{SCOD}) when primary treated effluent was considered for treatment in present SBR system. Average specific denitrification rate (q_{DN}) varied from 4.64 to 5.42 mg of N/gm MLVSS. hr for primary treated slaughterhouse wastewater for 3 hr anoxic period. The average 4.0 hourly specific denitrification rate (q_{DN}) varied from 4.95 to 5.88 mg of N/gm MLVSS. hr. The previous rate of specific denitrification rate (q_{DN}) was found to be followed in similar results as reported by Barnes and Bliss [43].

4.4 CONCLUSIONS

The present experimental investigation demonstrated that sequential batch reactor (SBR) is a variable and efficient biological method to treat slaughterhouse wastewater in a single unit. The total react period of 8 hr (4 hr aerobic and 4 hr anoxic) yielded optimum carbon oxidation, nitrification, and denitrification for treatment of carbonaceous and nitrogenous wastewater. The increase in MLVSS level in the reactor exhibited the growth favoring environment of the microorganism. The pH level in the SBR descended initially during aerobic period due to nitrification and carbon oxidation followed by an increasing trend indicating the existence of denitrifiers. This phenomenon has also been established by the variation of alkalinity level during aerobic and anoxic react period. The estimated values of biokinetic coefficients (k, K_s, Y, k_d) showed reasonable agreement with the literature values. The kinetic data and rate reaction constants could be used for the design of a field scale SBR for treating slaughterhouse wastewater. A design rationale can be evaluated on the basis of present experimental data for the purpose of application of this technology in similar plants. The outcome of the present investigation results would be helpful for making a design rationale for SBR treatment of slaughterhouse wastewater and a pilot plant study can be conducted with real-life wastewater sample by application of derived data of present study. In the future scope of the study, microbial genomics study including phosphate removal aspects would be also considered. The influence of solid retention time (SRT) should be explored also. A real-time kinetics profile with automatic data plotting could be derived for explaining the process in more rational way. It is also suggested that optimization of the process and operation variable may be examined with soft computing tools using various statistical approach.

TABLE 5: Denitrification rates during anoxic react phase for treatment of slaughterhouse wastewater in SBR.

Initial NH$_4^+$-N (mg/L as N)	Initial SCOD (mg/L)	React period combination (Aerobic/Anoxic)	Avg. anoxic SCOD utilization rate (q$_{SCOD}$) (mg SCOD/gm MLVSS. hr)	Specific denitrification rate (q$_{DN}$) (mg N/gm MLVSS. hr)							
				1.0hr	2.0hr	3.0hr	4.0hr	5.0hr	Avg. (3.0hrly)	Avg. (4.0hrly)	Avg. (5.0hrly)
185.24	1028.55	(4+4)	26.25	4.49	5.57	5.85	3.89	—	5.30	4.95	—
173.88	1023.22	(5+3)	34.87	4.27	5.51	4.16	—	—	4.64	—	—
176.85	1042.52	(3+5)	38.15	4.57	5.55	6.16	7.24	6.23	5.42	5.88	5.95

REFERENCES

1. A. O. Akinro, I. B. Ologunagba, and O. Yahaya, "Environmental implications of unhygienic operation of a city abattoir in Akure, Western Nigeria," ARPN Journal of Engineering and Applied Sciences, vol. 4, no. 9, pp. 311–315, 2009.
2. V. P. Singh and S. Neelam, "A survey report on impact of abattoir activities management on environments," Indian Journal of Veterinarians, vol. 6, pp. 973–978, 2011.
3. S. M. Gauri, "Treatment of wastewater from abattoirs before land application: a review," Bioresource Technology, vol. 97, no. 9, pp. 1119–1135, 2006.
4. Y. O. Bello and D. T. A. Oyedemi, "Impact of abattoir activities and management in residential neighbourhoods: a case study of Ogbomoso, Nigeria," Journal of Social Science, vol. 19, pp. 121–127, 2009.
5. D. Muhirwa, I. Nhapi, U. Wali, N. Banadda, J. Kashaigili, and R. Kimwaga, "Characterization of wastewater from an abattoir in Rwanda and the impact on downstream water quality," International Journal of Ecology, Development, vol. 16, no. 10, pp. 30–46, 2010.
6. A. O. Aniebo, S. N. Wekhe, and I. C. Okoli, "Abattoir blood waste generation in rivers state and its environmental implications in the Niger Delta," Toxicological and Environmental Chemistry, vol. 91, no. 4, pp. 619–625, 2009.
7. O. Chukwu, "Analysis of groundwater pollution from abattoir Waste in Minna, Nigeria," Research Journal of Diary Science, vol. 2, pp. 74–77, 2008.
8. J. Grady, G. Daigger, and H. Lim, Biological Wastewater Treatment, Marcel Dekker, New York, NY, USA, 1999.
9. I. Ruiz, M. C. Veiga, P. de Santiago, and R. Blfizquez, "Treatment of slaughterhouse wastewater in a UASB reactor and an anaerobic filter," Bioresource Technology, vol. 60, no. 3, pp. 251–258, 1997.
10. D. I. Massé and L. Masse, "Treatment of slaughterhouse wastewater in anaerobic sequencing batch reactors," Canadian Agricultural Engineering, vol. 42, no. 3, pp. 131–137, 2000.
11. K. Sombatsompop, A. Songpim, S. Reabroi, and P. Inkong-ngam, "A comparative study of sequencing batch reactor and movingbed sequencing batch reactor for piggery wastewater treatment," Maejo International Journal of Science and Technology, vol. 5, no. 2, pp. 191–203, 2011.
12. Y. Rahimi, A. Torabian, N. Mehrdadi, and B. Shahmoradi, "Simultaneous nitrification-denitrification and phosphorus removal in a fixed bed sequencing batch reactor (FBSBR)," Journal of Hazardous Materials, vol. 185, no. 2-3, pp. 852–857, 2011.
13. E. Bazrafshan, F. K. Mostafapour, M. Farzadkia, K. A. Ownagh, and A. H. Mahvi, "Slaughterhouse wastewater treatment by combined chemical coagulation and electrocoagulation process," PLOS ONE, vol. 7, no. 6, 2012.
14. R. Rajakumar, T. Meenambal, P. M. Saravanan, and P. Ananthanarayanan, "Treatment of poultry slaughterhouse wastewater in hybrid upflow anaerobic sludge blanket reactor packed with pleated poly vinyl chloride rings," Bioresource Technology, vol. 103, no. 1, pp. 116–122, 2012.

15. G. C. Sunder and S. Satyanarayan, "Efficient treatment of slaughter house wastewater by anaerobic hybrid reactor packed with special floating media," International Journal of Chemical and Physical Sciences, vol. 2, pp. 73–81, 2013.

16. European Commission, Integrated Pollution Prevention and Control: Reference Document on Best Available Techniques in the Slaughterhouses and Animal by-Products Industries, 2005.

17. A. H. Mahvi, "Sequencing batch reactor: a promising technology in wastewater treatment," Iranian Journal of Environmental Health Science and Engineering, vol. 5, no. 2, pp. 79–90, 2008.

18. N. Z. Al-Mutairi, M. F. Hamoda, and I. A. Al-Ghusain, "Slaughterhouse wastewater treatment using date seeds as adsorbent," Journal of Environment Science and Health, vol. 55, pp. 678–710, 2007.

19. R. Boopathy, C. Bonvillain, Q. Fontenot, and M. Kilgen, "Biological treatment of low-salinity shrimp aquaculture wastewater using sequencing batch reactor," International Biodeterioration and Biodegradation, vol. 59, no. 1, pp. 16–19, 2007.

20. H. S. Kim, Y. K. Choung, S. J. Ahn, and H. S. Oh, "Enhancing nitrogen removal of piggery wastewater by membrane bioreactor combined with nitrification reactor," Desalination, vol. 223, no. 1–3, pp. 194–204, 2008.

21. N. Z. Al-Mutairi, F. A. Al-Sharifi, and S. B. Al-Shammari, "Evaluation study of a slaughterhouse wastewater treatment plant including contact-assisted activated sludge and DAF," Desalination, vol. 225, no. 1–3, pp. 167–175, 2008.

22. D. Roy, K. Hassan, and R. Boopathy, "Effect of carbon to nitrogen (C:N) ratio on nitrogen removal from shrimp production waste water using sequencing batch reactor," Journal of Industrial Microbiology and Biotechnology, vol. 37, no. 10, pp. 1105–1110, 2010.

23. R. Rajagopal, N. Rousseau, N. Bernet, and F. Béline, "Combined anaerobic and activated sludge anoxic/oxic treatment for piggery wastewater," Bioresource Technology, vol. 102, no. 3, pp. 2185–2192, 2011.

24. J. Palatsi, M. Vinas, M. Guivernau, B. Fernandez, and X. Flotats, "Anaerobic digestion of slaughterhouse waste: main process limitations and microbial community interactions," Bioresource Technology, vol. 102, no. 3, pp. 2219–2227, 2011.

25. C. Kern and R. Boopathy, "Use of sequencing batch reactor in the treatment of shrimp aquaculture wastewater," Journal of Water Sustainability, vol. 2, no. 4, pp. 221–232, 2012.

26. F. Wang, Y. Liu, J. Wang, Y. Zhang, and H. Yang, "Influence of growth manner on nitrifying bacterial communities and nitrification kinetics in three lab-scale bioreactors," Journal of Industrial Microbiology and Biotechnology, vol. 39, no. 4, pp. 595–604, 2012.

27. D. Obaja, S. Mac, and J. Mata-Alvarez, "Biological nutrient removal by a sequencing batch reactor (SBR) using an internal organic carbon source in digested piggery wastewater," Bioresource Technology, vol. 96, no. 1, pp. 7–14, 2005.

28. K. V. Lo and P. H. Liao, "A full-scale sequencing batch reactor system for swine wastewater treatment," Journal of Environmental Science and Health, vol. 42, no. 2, pp. 237–240, 2007.

29. A. H. Mahvi, A. R. Mesdaghinia, and F. Karakani, "Nitrogen removal from waste-water in a continuous flow sequencing batch reactor," Pakistan Journal of Biological Sciences, vol. 7, no. 11, pp. 1880–1883, 2004.

30. R. Lemaire, Z. Yuan, B. Nicolas, M. Marcelino, G. Yilmaz, and J. Keller, "A se-quencing batch reactor system for high-level biological nitrogen and phosphorus re-moval from abattoir wastewater," Biodegradation, vol. 20, no. 3, pp. 339–350, 2009.

31. S. Q. Aziz, H. A. Aziz, M. S. Yusoff, and M. J. K. Bashir, "Landfill leachate treat-ment using powdered activated carbon augmented sequencing batch reactor (SBR) process: optimization by response surface methodology," Journal of Hazardous Ma-terials, vol. 189, no. 1-2, pp. 404–413, 2011.

32. G. Durai, N. Rajamohan, C. Karthikeyan, and M. Rajasimman, "Kinetics studies on biological treatment of tannery wastewater using mixed culture," International Journal of Chemical and Biological Engineering, vol. 3, no. 2, pp. 105–109, 2010.

33. M. Faouzi, M. Merzouki, and M. Benlemlih, "Contribution to optimize the biologi-cal treatment of synthetic tannery effluent by the sequencing batch reactor," Journal of Materials and Environmental Science, vol. 4, no. 4, pp. 532–541, 2013.

34. S. Dey and S. Mukherjee, "Performance and kinetic evaluation of phenol biodegra-dation by mixed microbial culture in a batch reactor," International Journal of Water Resources and Environmental Engineering, vol. 2, no. 3, pp. 40–49, 2010.

35. American Public Health Association, American Water Works Association, Water Pollution Control Federation, Standard Methods for the Examination of Water and Wastewater, American Water Works Association, Washington, DC, USA, 20th edi-tion, 1998.

36. H. Furumai, A. Kazmi, Y. Furuya, and K. Sasaki, "Modeling long term nutrient re-moval in a sequencing batch reactor," Water Research, vol. 33, no. 11, pp. 2708–2714, 1999.

37. A. Tremblay, R. D. Tyagi, and R. Y. Surampalli, "Effect of SRT on nutrient removal in SBR system," Practice Periodical of Hazardous, Toxic, and Radioactive Waste Management, vol. 3, no. 4, pp. 183–190, 1999.

38. A. W. Lawrence and P. L. McCarty, "Unified basis for biological treatment design and operation," Journal of the Sanitary Engineering Division, vol. 96, no. 3, pp. 757–778, 1970.

39. J. P. Li, M. G. Healy, X. M. Zhan, and M. Rodgers, "Nutrient removal from slaugh-terhouse wastewater in an intermittently aerated sequencing batch reactor," Biore-source Technology, vol. 99, no. 16, pp. 7644–7650, 2008.

40. P. Fongsatitkul, D. G. Wareham, P. Elefsiniotis, and P. Charoensuk, "Treatment of a slaughterhouse wastewater: effect of internal recycle rate on chemical oxygen de-mand, total Kjeldahl nitrogen and total phosphorus removal," Environmental Tech-nology, vol. 32, no. 15, pp. 1755–1759, 2011.

41. R. Kanimozhi and N. Vasudevan, "Effect of organic loading rate on the performance of aerobic SBR treating anaerobically digested distillery wastewater," Clean Tech-nologies and Environmental Policy, vol. 15, pp. 511–528, 2013.

42. D. Kulikowska and E. Klimiuk, "Removal of organics and nitrogen from municipal landfill leachate in two-stage SBR reactors," Polish Journal of Environmental Stud-ies, vol. 13, no. 4, pp. 389–396, 2004.

43. J. Doyle, S. Watts, D. Solley, and J. Keller, "Exceptionally high-rate nitrification in sequencing batch reactors treating high ammonia landfill leachate," Water Science and Technology, vol. 43, no. 3, pp. 315–322, 2001.
44. Metcalf and Eddy Inc., Wastewater Engineering Treatment Disposal Reuse, Tata McGraw-Hill, 3rd edition, 1995.
45. D. P. Barnes and P. J. Bliss, Biological Control of Nitrogen in Wastewater Treatment, E. & F.N. Spon, London, UK, 1983.

CHAPTER 5

Physicochemical Process for Fish Processing Wastewater

NEENA SUNNY AND LEKHA MATHAI

5.1 INTRODUCTION

Fisheries play a significant role in the economic and social wellbeing of nations. Fisheries and its resources constitute a wide source of food and feed a large part of the world's population besides being a huge employment sector. The fish industry consumes a large amount of water in operations such as cleaning, washing, cooling, thawing, ice removal, etc. Consequently, this sector also generates large quantities of wastewater in which the treatment is particularly difficult due to the high content of organic matter and salts and to the significant amount of oil and grease they present. These factors, together with the fact that these effluents present significant variations depending on the production process and on raw material processed makes difficult to meet the emission limit values for industrial wastewaters and to deal with this problem in a sustainable manner. A wide variety of inhibitory substances are the primary cause of anaerobic digester upset or failure since they are present in substantial

Physicochemical Process for Fish Processing Wastewater. Sunny N and Lekha Mathai L. International Journal of Innovative Research in Science, Engineering and Technology *2,4 (2013). Reprinted with permission from the* Journal of Innovative Research in Science, Engineering and Technology.

concentrations in wastes. These effluents are often subjected to a pre-treatment before discharge to the sewage system for further treatment at an urban wastewater treatment plant. The common processes in fish processing plants are filleting, freezing, drying, fermenting, canning and smoking (Palenzuela-Rollon, 1999).Regarding the organic matter degradation, the wastewaters are conventionally submitted to biological treatments. It yields desired results only when carried out in optimum conditions under proper observation. This paper presents a critical review of the inhibitors that affect the wastewater treatment in fish processing industry.

5.2 IMPORTANCE OF PHYSICOCHEMICAL PROCESS

The inhibitors commonly present in the fish processing wastewater include wastewater salinity, fat, oil and grease, ammonia content system pH and high content of organic matter.

Hypersaline effluents are generated by various industrial activities such as seafood processing, vegetable canning, and pickling, tanning and chemical manufacturing. This wastewater, rich in both organic matter and total dissolved solids, is difficult to treat using conventional biological wastewater treatment processes (Ludzack and Noran, 1965). Saline wastewater biological treatment systems usually result in low BOD removal because of the adverse effects of salt on the microbial flora. High salt concentrations cause plasmolysis or loss of activity of cells. So an efficient treatment process for these saline wastewaters has to be considered since conventional wastewater treatment processes will not give better results with saline wastewater. Lipids (characterized as oils, grease, fats and long-chain fatty acids) are important organic components of wastewater. Carawan et al.(1979) reported the FOG values foe herring, tuna, salmon and catfish processing were 60–800mg/l, 250mg/l.20–550mg/l and 200mg/l respectively. The behavior of lipids in biological treatment systems has led to many studies, which have evaluated their removal, but still the exact behavior of lipids in these processes is not well understood. The main components of fish processing wastewater are lipids and protein (Gonzalez, 1996). The Fat, oil and grease (FOG) should be removed from wastewater because it usually floats on the water's surface and affects the oxygen transfer to water.

TABLE 1: Performance of aerobic and anaerobic systems for fish processing wastewater treatment

Process	Wastewater Characteristics (mg/l)	Organic loading	Removal Efficiency	Reference
Aerobic				
Activated sludge	–	0.5 kg BOD/ m^3d	90–95% BOD	Carawan et al. (1979)
Aerated lagoon	–	–	90–95% BOD	Carawan et al. (1979)
Trickling filter	BOD up to 3000	0.08–0.4 kg BOD/m^3d	80–87% BOD	Park et al. (2001)
Rotating biological contactor	pH 6–7; COD 6000–9000; BOD 5100; TSS 2000; TKN 750	0.018–0.037 kg COD/m^3d	85–98% COD	Najafpour et al. (2006)
Anaerobic				
Anaerobic fluidized bed reactor	COD 90000;	6.7 kg COD/ m^3d	88% COD	Balslev-Olesen et al. (1990)
Anaerobic fixed film	–	2 kg COD/m^3d	75% COD	Veiga et al. (1991)
Anaerobic digester	COD 34,500; TS 4000l Cl⁻ 14,000	4.5 kg COD/ m^3d	80% COD	Mendex et al. (1992)
Anaerobic filter	–	0.3–0.99 kg COD/m^3d	78–84% COD	Prasertsan et al. (1994)
Upflow anaerobic sludge blanket reactor	COD 2718 ± 532; lipids 232 ± 29; TKN 410 ± 89; pH 7.2–7.6	1–8 kg COD/ m^3 d	80–95% COD	Palenzuela-Rollon et al. (2002)

Ammonia is a common hydrolysis product during waste degradation and could cause inhibition when present in high concentrations. pH could be an inhibition factor alone and interacts with ammonia inhibition by changing the relative ratio of ammonium to ammonia. Ammonia emission and proteinaceous matter decomposition is mostly pH dependent (Gonzalez, 1996). The high nitrogen levels are likely due to the high protein content (15–20% of wet weight) of fish and marine invertebrate (Sikorski, 1990). Sometimes high ammonia concentration is observed due to high blood and slime content in wastewater streams. As reported by a few fish processing plant the overall, ammonia concentration ranged from 0.7 mg/L to 69.7 mg/L (Technical Report Series FREMP, 1994). In the fish condensate the total ammonia content can be up to approximately 2000 mg N/L. High BOD concentrations are generally associated with high ammonia concentrations (Technical Report Series FREMP, 1994). The degree of ammonia toxicity depends primarily on the total ammonia concentration and pH.

Wastewater loading rate is a critical design factor for wastewater treatment systems. Organic loading rate is summarized in Table 1. In anaerobic wastewater treatment, loading rate plays an important role. In the case of nonattached biomass reactors, where the hydraulic retention time is long, overloading results in biomass washout. This, in turn, leads to process failure. Fixed film, expanded and fluidized bed reactors can withstand higher organic loading rate.

5.3 BIOLOGICAL TREATMENT PROCESSES

5.3.1 EFFECT OF SALINITY

Fish processing industries require a large amount of salt for fish conservation. High salinity of wastewater strongly inhibits the aerobic biological treatment of wastewater. Stewart et al. (1962) reported considerable BOD5 reduction due to the combined effect of high salinity and high organic loading. Kincannon and Gaudy (1968) observed that due to rapid change in salinity soluble COD was increased by the release of cellular material. They found that relatively more oxygen was used by cells grown in the presence of high salt concentrations. High saline concentrations have negative effects on organic matter and nitrogen removal (Intrasungkha et al., 1999). Aloui et al., 2009 proposed an activated sludge process for the treatment of fish processing saline water. However, the pollution abatement rates decreased with increasing the COD loading rate and salt content. Inhibition process was found to be significant for salt concentrations higher than 4% NaCl.

It is well known that anaerobic treatment of wastewater is inhibited by the presence of high sodium/or chloride concentrations. Methanogenesis is strongly inhibited by a sodium concentration of more than 10 g/L (Lefebvre and Moletta, 2006). But the work of Omil et al. (1995) on fish-processing effluent using an anaerobic contact system showed that the adaptation of an active methanogenic biomass at the salinity level of the effluent was possible with a suitable strategy. Acclimation of methanogens to high concentrations of sodium over prolonged periods of time could increase the tolerance and shorten the lag phase before methane production begins (de Baere et al., 1984; Feijoo et al., 1995; Omil et al., 1995a,b, 1996b; Chen et al., 2003). The tolerance is related to the Na^+ concentration the methanogens acclimated to and the time of exposure. Anaerobic digesters are usually more sensitive to high salinity than an activated sludge unit.

5.3.2 EFFECT OF LIPIDS

Fish lipid contains long-chain n-3 (omega-3) PUFA, particularly EPA (C20:5 n-3) and DHA (C22:6 n-3). In aerobic wastewater treatment systems, lipids are generally believed to be biodegradable and, therefore, considered as part of the organic load that is treated. However, lipids have detrimental effects on oxygen transfer. They reduce the rates at which oxygen is transferred, thereby depriving the microorganisms of oxygen. This effect results in reduced microbial activity. To enhance biodegradation of lipids, Keenan and Sabelnikov (2000) proposed the use of a combination of suspended and attached growth treatment systems. They found that the lipid content in the effluent wastewater could not be reduced to values below 0.3 g/l from 1.512 g/l by using a suspended growth treatment system only, whereas adding a biofilter (a solid support that could be colonized by bacteria) to the suspended growth system substantially reduced the lipid content in the wastewater effluent to 0.028 g/l. However, the treatment system reported by Keenan and Sabelnikov (2000) sporadically failed, and the content of lipids in the effluent wastewater increased to 0.386 g/l. Although the authors attributed the sporadic failures to the failure of the pH adjustment system, the complete explanation for such failures was unknown.

The treatment of lipid-rich wastewater is still a challenge. In addition to aerobic wastewater treatment systems, anaerobic systems are also widely used for treatment of lipid-rich wastewater. Most importantly, high-rate anaerobic treatment systems have been developed. Among these systems, the upflow anaerobic sludge bed (UASB) reactor is the most widely used in the treatment of domestic and industrial wastewater due to its low-cost and adequate treatment efficiency. Although UASB reactors have been well characterized, and their usefulness for treatment of municipal and Industrial wastes well documented their treatment fallures have also been reported when treating lipid-rich wastewater. Gujer and Zehnder(1983) demonstrated that low density of the floating aggregates slows the biodegradation of lipids.

In order to improve lipid biodegradation in such troublesome systems, Rinzema (1994) proposed rigorous mixing as a means of maintaining good contact between bacteria and lipids in the anaerobic digester. In this regard, Li et al. (2002) proposed a two-stage anaerobic digestion process consisting of a mixing unit and a high solids digestion unit for treatment of lipid-rich wastewater. However, the degradation efficiency decreased at loading rates above 20 and 33 kg COD/m^3 day under mesophilic (35⁰C) and thermophilic (55⁰C) conditions, respectively. At higher loading rates, low degradation of lipids is expected. To solve this problem, Van Lier et al. (1994) introduced a new concept of multi-stage UASB reactor, which consists of a number of gas–solids separators. Further, Tagawa et al. (2002) investigated the ability of a multi-stage UASB reactor under thermophilic conditions (55⁰C) to treat lipid-rich wastewater at retention times from 0 to 600 days. But the overall COD removal (based on the total effluent COD) was very unsatisfactory at only 60–70%. Lettinga carried out further modifications of the UASB reactor (so-called expanded granular sludge bed reactor, EGSB). Rinzema et al. (1994) took advantage of the EGSB reactors and observed no flotation of granular sludge, and achieved a high volumetric loading rate of 31.4 g COD/l day. Hence, when treating lipid-rich wastewater, it should be advantageous to run sequencing cycles of adsorption and degradation in order to enhance complete removal of lipids. This lipid considered as the highly attention source for human consumption as well as industrial use. In this sense, the financial benefits can be obtained and environmental pollution is certainly decreased.

5.3.3 EFFECT OF PH AND AMMONIA

According to Boone and Xun (1987) most methanogenic bacteria have optima for growth between pH 7 and 8, whereas VFA degrading bacteria have lower pH optima. The optimal pH for mesophilic biogas reactor is 6.7–7.4 (Clark and Speece, 1971). The study of Sandberg and Ahring (1992) demonstrated that fish condensate can be treated well in a UASB reactor from pH 7.3 to 8.2. When the pH was increased slowly to 8.0 or more, COD removal drop about 15–17%. Acetate was the only carbon source in the condensate that accumulated upon increasing the pH. It was concluded that gradual pH increment was essential in order to achieve the necessary acclimatization of the granules and to prevent disintegration of the granules and that the pH should not exceed 8.2. Aspe et al. (2001) modeled the ammonia-induced inhibition phenomenon of anaerobic digestion and concluded that methanogenesis was the most inhibited stage. The methanogenic activity was reduced by the presence of high concentrations of ammonia as a result of protein degradation during the anaerobic treatment. Ammonia inhibition was directly related to the concentration of the undissociated form (NH_3), therefore being more important at high pH levels. It was also reported that free ammonia (FA) inhibitory concentrations for mesophilic treatment have been 25–140 mg N-FA/L whereas during the thermophilic digestion of cattle manure, higher values, 390–700 mg N-FA/L, were tolerated after an initial acclimation period (Guerrero et al., 1997). Control of pH within the growth optimum of microorganisms may reduce ammonia toxicity (Bhattacharya and Parkin, 1989).The traditional method for removal of ammonia and organic pollutants from wastewater is biological treatment but ion exchange offers a number of advantages including the ability to handle shock loading and the ability to operate over a wider range of temperatures.

5.4 CONCLUSIONS

Wastewater characterization is a critical factor in establishing a corresponding effective management strategy or treatment process. Based on the studies conducted by different authors regarding the wastewater treatment in fish processing industry, which includes both aerobic and anaerobic treatment processes the following conclusions could be drawn. Anaerobic treatment processes are most widely used for treating wastewaters but these processes partly degrade wastewater containing fats and nutrients.

So, subsequent treatment is necessary for wastewater. The fats and nutrients could easily be removed in aerobic reactors. But a high-energy requirement by aerobic treatment methods is the primary drawback of these processes. In order to reduce energy consumption in aerobic treatment, physico chemical treatment processes may be combined with anaerobic-aerobic combination. An integrated design using physicochemical process followed by biological process would yield better treatment efficiency with less energy consumption and reduced sludge production. Combining physicochemical adsorptive treatment with biological treatment can provide synergistic benefits to the overall removal processes. Ion exchange removal solves some of the common operational reliability limitations of biological treatment, like slow response to environmental changes and leaching. Biological activity can in turn help reduce the economic and environmental challenges of ion exchange processes, like regenerant cost and brine disposal.

REFERENCES

1. Aspe, E., Marti, M.C., Jara, A., Roeckel, M., Ammonia inhibition in the anaerobic treatment of fishery effluents. Water Environ. Res. 73 (2), 154–164, 2001.
2. Balslev-Olesen, P., Lynggaard-Jensen, A., Nickelsen, C., Pilot-scale experiments on anaerobic treatment of wastewater from a fish processing plant. Water Sci. Technol. 22 (1–2), 463–474, 1990.
3. Boone, D.R., Xun, L., Effects of pH, temperature and nutrients on propionate degradation by a methanogenic enrichment culture. Appl. Environ. Microbiol. 53, 1589–1592, 1987.
4. Carawan, R.E., Chambers, J.V., Zall, J.V., Seafood water and wastewater management. North Carolina Agricultural Extension Services, Raleigh, NC, 1979.
5. Chen, W.H., Han, S.K., Sung, S., Sodium inhibition of thermophilic methanogens. J. Environ. Eng. 129 (6), 506–512, 2003.
6. Clark, R.H., Speece, R.E., The pH tolerance of anaerobic digestion. Adv. Water Pollut. Res. 1, 1–14, 1971.
7. de Baere, L.A., Devocht, M., van Assche, P., Verstraete, W., Influence of high NaCl and NH4Cl salt levels on methanogenic associations. Water Res. 18, 543–548, 1984.
8. Fathi Aloui, Sonia Khoufi, Slim Loukil, Sami Sayadi., Performances of an activated sludge process for the treatment of fish processing saline wastewater. Desalination 248 , 68–75, 2009.
9. Feijoo, G., Soto, M., Mende´z, R., Lema, J.M., Sodium inhibition in the anaerobic digestion process: antagonism and adaptation phenomena. Enzyme Microb. Technol. 17, 180–188, 1995.

10. Gonzalez, j.f. Wastewater treatment in the fishery industry. FAO Fisheries technical paper (FAO), no. 355/FAO, Rome (Italy), Fisheries, 1996.

11. Guerrero, L., Omil, F., Mthdez, R., Lema, J.M., (1997). Treatment of saline wastewaters from fish meal factories in an anaerobic filter under extreme ammoniacal concentrations. Bioresour. Technol. 61, 69–78, 1997.

12. Gujer W, Zehnder AJB Conversion process in anaerobic digestion. Water Sci Technol 15:127–167deptSikorski, Z., 1990. Seafood Resources: Nutrient Composition and Preservation. CRC Press Inc., Boca Raton, 1983.

13. Hsu TC, Hanaki K, Matsumoto J Kinetics of hydrolysis, oxidation and adsorption during olive oil degradation by activated sludge. Biotechnol Bioeng 25:1829–1839,1983.

14. Intrasungkha, N., Keller, J., Blackall, L.L., Biological nutrient removal efficiency in treatment of saline wastewater. Water Sci. Technol. 39 (6), 183–190, 1999.

15. Keenan D, Sabelnikov A, Biological augmentation eliminates grease and oil in bakery wastewater. Water Environ Res, 72: 141–146, 2000.

16. Kincannon, D.F., Gaudy, A.F., Response of biological waste treatment systems to changes in salt concentrations. Biotechnol. Bioeng. 10, 483– 496, 2006.

17. Lefebvre, O., Moletta, R., Treatment of organic pollution in industrial saline wastewater: a review. Water Res. 40, 3671–3682, 1968.

18. Lettinga, G., Hobma, S.W., Klapwijk, A., Van Velsen, A.F.M., de Zeeuw, W.J., Use of the Upflow Sludge Blanket (USB) reactor concept for biological wastewater treatment. Biotechnol. Bioeng. 22, 699–734, 1980.

19. Li YY, Sasaki H, Yamashita K, Seki K, Kamigochi I High-rate methane fermentation of lipid-rich food wastes by a high-solids co-digestion process. Water Sci Technol 45:143– 150, 2002.

20. Ludzack, F.J., Noran, P.K., Tolerance of high salinities by conventional wastewater treatment process. J. Water Pollut. Control Fed. 37 (10), 1404–1416, 1965.

21. Mendez, R., Omil, F., Soto, M., Lema, J.M., Pilot plant studies on the anaerobic treatment of different wastewater from a fish-canning factory. Water Sci. Technol. 25 (1), 37–44, 1992.

22. Najafpour, G.D., Zinatizadeh, A.A.L., Lee, L.K., Performance of a three-stage aerobic RBC reactor in food canning wastewater treatment. Biochem. Eng. J. 30, 297–302,2006.

23. Omil, F., Mende´z, R., Lema, J.M., Characterization of biomass from a pilot plant digester treating saline wastewater. J. Chem. Tech. Biotechnol. 63, 384–392,1995a.

24. Omil, F., Mende´z, R., Lema, J.M., Anaerobic treatment of saline wastewaters under high sulphide and ammonia content. Bioresour. Technol. 54, 269–278,1995b.

25. Palenzuela-Rollon, A , Anaerobic Digestion of Fish Processing Wastewater with Special Emphasis on Hydrolysis of Suspended Solids. Taylor and Francis, London, 1999.

26. Pankaj Chowdhury, T. Viraraghavan , A. Srinivasan. Biological treatment processes for fish processing wastewater – A review. Bioresource Technology 101: 439–449, 2010.

27. Park, E., Enander, R., Barnett, M.S., Lee, C., Pollution prevention and biochemical oxygen demand reduction in a squid processing facility. J. Cleaner Product. 9, 341–349, 2001.

28. Prasertsan, P., Jung, S., Buckle, K.A., Anaerobic filter treatment of fishery wastewater. World J. Microbiol. Biotechnol. 10, 11–13, 1994.

29. Rinzema A Anaerobic digestion of long-chain fatty acids in UASB and expanded granular sludge bed reactors. Proc Biochem 28:527–537, 1993.

30. Rinzema A, Alphenaar A, Lettinga G The effect of lauric acid shock loads on the biological and physical performance of granular sludge in UASB reactors digesting acetate. J Chem Technol Biotechnol 46:257–266, 1989.

31. Rinzema A, Boone M, van Knippenberg K, Lettinga G Bacterial effect of long-chain fatty acids in anaerobic digestion. Water Environ Res 66:40–49

32. Sandberg, M., Ahring, B.K., (1992). Anaerobic treatment of fish meal process wastewater in a UASB reactor at high pH. Appl. Microbiol. Biotechnol. 36, 800–804, 1994.

33. Sikorski, Z., Seafood Resources: Nutrient Composition and Preservation. CRC Press Inc., Boca Raton, 1990.

34. Stewart, M.J., Ludwig, H.F., Kearns, W.H., Effects of varying salinity on the extended aeration process. J. Water Pollut. Control Fed. 34, 1161–1177, 1962.

35. Tagawa T, Takahashi H, Sekiguchi Y, Ohashi A, Harada H Pilot-plant study on anaerobic treatment of a lipid- and protein-rich food industrial wastewater by a thermophilic multi-staged UASB reactor. Water Sci Technol 45:225–230, 2002.

36. Technical Report Series FREMP WQWM-93-10, DOE FRAP 1993-39, Wastewater Characterization of Fish Processing Plant Effluents. Fraser River Estuary Management Program. New West Minister, B. C,1994.

37. Van Lier JB, Boersma F, Debets MMWH, Lettinga G High-rate thermophilic anaerobic wastewater treatment in compartmentalized upflow reactors. Water Sci Technol 30:251–261,1994.

38. Veiga, M.C., Mendez, R.J., Lema, J.M., Treatment of tuna processing wastewater in laboratory and pilot scale DSFF anaerobic reactors. In: Proceedings of the 46th industrial waste conference, May 14–16, 1991. Purdue University, West Lafayett, Indiana, pp. 447–453,1991.

CHAPTER 6

Sustainable Agro-Food Industrial Wastewater Treatment Using High Rate Anaerobic Process

RAJINIKANTH RAJAGOPAL, NOORI M. CATA SAADY, MICHEL TORRIJOS, JOSEPH V. THANIKAL, AND YUNG-TSE HUNG

6.1 INTRODUCTION

Agro-industries are major contributors to worldwide industrial pollution. Effluents from many agro-food industries are a hazard to the environment and require appropriate and a comprehensive management approach. Worldwide, environmental regulatory authorities are setting strict criteria for discharge of wastewaters from industries. As regulations become stricter, there is now a need to treat and utilize these wastes quickly and efficiently. With the tremendous pace of development of sustainable biotechnology, substantial research has been devoted recently to cope with wastes of ever increasing complexity generated by agro-industries. Anaerobic digestion is an environmentally friendly green biotechnology to treat agro-food industrial effluents. In addition, the carbon emission and, therefore, the carbon footprint of water utilities is an important issue nowadays. In this perspective, it is essential to consider the prospects for the reduction of the carbon footprint from small and large wastewater treatment plants. The use of anaerobic treatment processes rather than aerobic would accomplish this purpose, because no aeration is required and the biogas generated can be used within the plant. Anaerobic digestion is unique, as it reduces waste and produces energy in the form of methane. Not only does this technology have a positive net energy production, but the biogas produced can also replace fossil fuel; therefore, it has a direct positive effect on greenhouse gas reduction. Thus, the carbon-negative anaerobic digestion process is considered as a sustainable wastewater treatment technology, which also provides the best affordable (low-cost) process for public health and environmental protection, as well as resource recovery. The attractiveness of biogas technology for large scale applications has been limited, essentially, because of the slow rate and process instability of anaerobic digestion. The slow rate means large digester volumes (and consequently, greater costs and space requirements) and process instability means the lack of assurance for a steady energy supply. These two major disadvantages of conventional anaerobic processes have been overcome by high rate anaerobic reactors, which employ cell immobilization techniques, such as granules and biofilms. Thus, various reactor designs that employ various ways of retaining biomass within the reactor have been developed over the past two decades. The purpose of this article is to summarize the current status of the research on high rate anaerobic treatment of agro-food industrial wastewater and to provide strategies to overcome some of the operational problems.

6.2 AGRO-FOOD INDUSTRIAL WASTEWATERS

About 65%–70% of the organic pollutants released in the water bodies in India are from food and agro-product industries, such as distilleries, sugar factories, dairies, fruit canning, meat processing and pulp and paper mills [1]. Similarly, the pulp and paper industry is one of the most significant industries in Sweden, as well as many other countries around the world, and the products constitute important industrial trade in terms of value of production [2]. Wine production is one of the leading agro-food industries in Mediterranean countries, and it has also attained importance in other parts of the world, such as Australia, Chile, the United States, South Africa and China, with increasing influence on the economy of these countries [3]. The wine industry generates huge volumes of wastewater that are mainly originated from several washing operations, e.g., during crushing and pressing of the grapes, cleaning the fermentation tanks, containers, other equipment and surfaces. In addition to this, olive oil industries have gained fundamental economic importance for many Mediterranean countries [4]. Malaysia presently accounts for 39% of world palm oil production and 44% of world exports [5]. Due to its surplus production, a huge amount of polluted wastewater, commonly referred to as palm oil mill effluent (POME) is generated. Fia et al. [6] reported that in coffee producing regions, such as Brazil, Vietnam and Colombia, the final effluent produced from this process has become a large environmental problem, creating the need for low cost technologies for the treatment of wastewater. Nieto et al. [7] determined the potential for methane production from six agro-food wastes (beverage waste, milled apple waste, milk waste, yogurt waste, fats and oils from dairy wastewater treatment and cattle manure). The wastewater generation varies from country to country. For instance, world wine production in 2011 was estimated to be about 270 million hectoliters, with the European Union producing 152 million hL (France with 50.2 million hL; Italy, with 40.3 million hL; Spain, with 35.4 million hL), whereas it was estimated to be 18.7, 15.5, 11.9, 10.6, 9.3 and 2.3 million hL for countries, such as the USA, Argentina, Australia, Chile, South Africa and New Zealand, respectively [8]. The annual worldwide production of olive oil is estimated to be about 1750 million metric tons, with Spain, Italy, Greece, Tunisia and Portugal being the major producers, and about 30 million cubic meters of oil mill wastes (OMW) are generated annually in the Mediterranean area during the seasonal extraction of olive oil [4,9].

The composition and concentration of different agro-food wastewaters vary from low (wash water from sugar mill or dairy effluents) to high strength substrates (cheese, winery and olive mill wastewaters), particularly in terms of organic matter, acids, proteins, aromatic compounds, available nutrients, etc. [1,3,10]. The main parameters of the agro-food industrial wastewater, such as total solids (TS), total nitrogen (TN), total

Efficient Management of Wastewater from Manufacturing

phosphorus (TP) and biochemical and chemical oxygen demand (BOD and COD), respectively are given in Table 1.

TABLE 1: Characteristics of typical agro-food industrial wastewater.

Industry	TS (mg L^{-1})	TP (mg L^{-1})	TN (mg L^{-1})	BOD (mg L^{-1})	COD (mg L^{-1})	Reference
Food processing[a]	-	3	50	600–4,000	1,000–8,000	[11]
Palm oil mill	40	-	750	25	50	[12]
Sugar-beet processing	6100	2.7	10	-	6,600	[13]
Dairy	1,100–1,600	-	-	800–1,000	1,400–2,500	[14]
Corn milling	650	125	174	3,000	4,850	[15]
Potato chips	5,000	100	250	5,000	6,000	[16]
Baker's yeast	600	3	275	-	6,100	[17]
Winery	150–200	40–60	310–410	-	18,000–21,000	[1,3]
Dairy	250–2,750	-	10–90	650–6,250	400–15,200	[18]
Cheese dairy	1,600–3,900	60–100	400–700	-	23,000–4,0000	[1]
Olive mill	75,500	-	460	-	130,100	[19]
Cassava starch	830	90	525	6,300	10,500	[20]

Notes: [a] contains flour, soybean, tomato, pepper and salt. TS: total solids; TN: total nitrogen; TP: total phosphorus; BOD: biochemical oxygen demand; COD: chemical oxygen demand.

6.3 HIGH RATE ANAEROBIC REACTORS

High-rate anaerobic digesters receive increasing interest, due to their high loading capacity and low sludge production [21]. The commonly used high-rate anaerobic digesters include: anaerobic filters, upflow anaerobic sludge blanket (UASB) reactors, anaerobic baffled, fluidized beds, expanded granular sludge beds (EGSB), sequencing batch reactors and anaerobic hybrid/hybrid upflow anaerobic sludge blanket reactors [1,21].

Rajagopal et al. [22] developed high-rate upflow anaerobic filters (UAFs) packed with low-density polyethylene media for the treatment of wastewater discharged from various agro-food industries with different composition and COD concentrations viz. synthetically prepared low strength (~1.9 g COD L^{-1}), fruit canning (~10 g COD L^{-1}), winery (~20 g COD L^{-1}) and cheese-dairy (~30 g COD L^{-1}) wastewaters. High organic loading rates (OLRs) (12–27 g COD L^{-1} d^{-1}) were reported in this study. Low-density polyethylene support (29 mm high; 30–35mm diameter; density: 0.93 kg m^{-3}; specific area: 320 m^2 m^{-3}) was able to retain between 0.7 and 1.6 g dried solids per support. This study concluded that the low-density polyethylene support is a good colonization matrix to increase the quantity of biomass in the reactor compared to conventional treatment systems.

In a similar study, Ganesh et al. [3] investigated the performance of upflow anaerobic fixed-bed reactors filled with low density supports of varying size and specific surface area for the treatment of winery wastewater. They found that efficiency of reactors increased with decrease in size and increase in specific surface area of the supports. A maximum OLR of 42 g L^{-1} d^{-1} with 80% COD removal efficiency was attained, when supports with the highest specific surface area were used. However, for long-term operation, clogging might occur in the reactors packed with small size supports.

Esparza et al. [23] reported that a pilot-scale upflow anaerobic sludge blanket (UASB) reactor treating cereal-processing wastewater at 17 °C with OLR 4–8 kg COD m^{-3} d^{-1} and a hydraulic retention time (HRT) of 5.2 h removed 82%–92% of the COD. Shastry et al. [24] investigated the feasibility of a UASB reactor system as a pretreatment for hydrogenated vegetable oil wastewater. COD removal efficiency of 99%–80% at OLR varying in the range 1.3–10 g COD L^{-1} d^{-1} was obtained with a specific methane yield of 0.30–0.35 m^3 CH_4 kg^{-1} COD.

Won and Lau [25] operated an anaerobic sequencing batch reactor (ASBR) to investigate the effect of pH, HRT and OLR on biohydrogen production at 28 °C. For a carbon-rich substrate, a maximum hydrogen production rate and yield of 3 L H_2 L^{-1} reactor d^{-1} and 2.2 mol H_2 mol^{-1} hexose, respectively, were achieved at pH 4.5, HRT 30 h and OLR 11.0

g L^{-1} d^{-1}. A mesophilic anaerobic sequencing batch biofilm reactor (AS-BBR) treating lipid-rich wastewater with OLR as much as 12,1 g COD L^{-1} d^{-1} was achieved with 90% COD removal efficiency [26].

Shanmugam and Akunna [27] investigated the performance of a granular bed baffled reactor (GRABBR) for the treatment of low-strength wastewaters at increasing OLRs (up to 60 g COD L^{-1} d^{-1}). They showed experimentally that GRABBR encouraged different stages of anaerobic digestion in separate vessels longitudinally across the reactor, and it had greater process stability at relatively short HRTs (1 h) with 86% methane in biogas. Bialek et al. [28] assessed the performance of two kinds of reactors (inverted fluidized bed (IFB) and expanded granular sludge bed (EGSB) reactors) treating simulated dairy wastewater. At 37 °C, they obtained more than 80% of COD and protein removals with an OLR of 167 mg COD L^{-1} h^{-1} and a HRT of 24 h.

Rajagopal et al. [29] proposed a modified UAF reactor design, called "hybrid upflow anaerobic sludge-filter bed (UASFB) reactor", to overcome problems faced by fixed bed reactors, such as clogging, short circuiting and biomass washout. This configuration contains a sludge bed in the lower part of the reactor and a filter bed in the upper part. For the treatment of wine distillery vinasse (21.7 g COD L^{-1}) they achieved a high OLR (18 g COD L^{-1} d^{-1}) at a short HRT (26 h), while maintaining high COD removal efficiencies of about 85%. However, an aerobic post-treatment is required to make the effluent fit for final disposal, especially in terms of nitrogen [3,14,22]. Table 2 summarizes the performance of anaerobic digesters for the treatment of various agro-food wastewaters in terms of design and applied operational conditions, process efficiency and energy characteristics.

6.3.1 OTHER TREATMENT STRATEGIES TO ENHANCE THE REACTOR PERFORMANCE

Biogas production can be augmented by using ample lignocellulose materials viz. agricultural and forest residues [30,31]. However, the complex lignocellulose structure limits the accessibility of sugars in cellulose and hemicellulose. Consequently, a pretreatment is essential and several potential pretreatment techniques have been developed for lignocellulose material [32].

Nkemka and Murto [31] evaluated the biogas production in batch and UASB reactors from pilot-scale acid catalyzed steam-pretreated and enzymatic-hydrolysed wheat straw. They showed that the pre-treatment increased the methane yield [0.28 m^3 kg^{-1} volatile solids (VS) fed] by 57% compared to untreated straw. The treatment of straw hydrolysate with nutrient supplementation in a UASB reactor resulted in a high methane production rate (2.70 m^3 m^{-3} d^{-1}) at an OLR of 10.4 g COD L^{-1} d^{-1} and with 94% COD reduction.

Badshah et al. [2] proved that the methanol condensate can be efficiently converted into biogas in a UASB reactor instead of methanol, with much of the smell of the feed eliminated. Bosco and Chiampo [33] reported polyhydroxyalkanoates (PHAs) (biodegradable plastic) production from milk whey and dairy wastewater activated sludge. They defined the suitable C/N ratio, the pre-treatments required to lower the protein content and the effect of pH correction.

Garcia et al. [34] investigated the effect of polyacrylamide (PAM) for the biomass retention in an UASB reactor treating liquid fraction of dairy manure at several organic loading rates. They have concluded that PAM addition enhanced sludge retention and reactor performance (with a total COD removal of 83% compared to 77% removal efficiencies for UASB without polymer addition).

In order to remove ammonia nitrogen, Yu et al. [35] attempted to enrich anammox bacteria in sequencing batch biofilm reactors (SBBRs) with different inoculations. The maximum total nitrogen loading rate of SBBR gradually reached 1.62 kg N m^{-3} d^{-1}, with a removal efficiency higher than 88%. Few researchers [36,37] reported that anammox bacteria have slow growth rate kinetics, and hence, they are vulnerable to external conditions, such as low temperature, high dissolved oxygen and some inhibitors. Although this technique faces some complications in full-scale applications, such as long start-up and instability, anammox is still an innovative technological development in reducing ammonia from wastewater [35,38].

TABLE 2: Performance of anaerobic digesters for the treatment of various agro-food wastewaters.

Substrate	Reactor type[a]	TCOD (g L⁻¹)	pH	HRT (h)	OLR (g COD L⁻¹ d⁻¹)	TCOD removal (%)	SCOD removal (%)	VFA/ alkalinity	Comments	Reference
Fruit canning	UAF	9–11.6	6–6.5	12	19	68	81	0.35	Specific sludge loading rate (SSLR): 0.56 g COD g⁻¹ VSS d⁻¹	[1]
Cheese dairy	UAF	23–40	6–6.5	40	17	71	82	0.6	SSLR: 0.63 g COD g⁻¹ VSS d⁻¹	[3]
Winery wastewater	FBR	18–21	8–11	15–23	22–42	65–70	80	0.8	SSLR: 0.93–1.2 g COD g⁻¹ VSS d⁻¹	[3]
Dairy manure	UASB	16.5–20.4	7.4	36–48	9–12.7	76.5–83.4	-	-	Methane yield: 0.296–0.312 L CH₄ g⁻¹ COD; Higher values corresponded to UASB with polymer addition	[34]
Dairy wastewater	AF/ BAF	1.8–2.4	6.6–8.4	18.2–38.6	0.66–0.72	98–99 (Overall)	-	-	Recirculation ratio ranged 100%–300%; Specific growth rate of the integrated biomass: 0.621–1.208 d⁻¹	[39]
Distillery vinasse	UAS-FB	3.1–21.7	6–6.5	25–27	17.9–18.2	80–82	84–87	<0.4	SSLR: 0.43–0.47 g COD g⁻¹ VS d⁻¹	[29]

TABLE 2: *Cont.*

Substrate	Reactor type[a]	TCOD (g L⁻¹)	pH	HRT (h)	OLR (g COD L⁻¹ d⁻¹)	TCOD removal (%)	SCOD removal (%)	VFA/ alkalinity	Comments	Reference
Food process-ing	JBl-LAFB	0.96–7.9	3.4–11.2	24	1.6–5.6	80	-	0.2–0.5	Methane production: 0.414 m³ L⁻¹ reactor volume day⁻¹; Specific energy input for the anaerobic reactor: 0.12 kWh m⁻³	[11]
Cassava starch wastewater	UMAR	10.5	4.5–4.92	6	10.2–40	77.5–92	-	-	The optimum HRT was 6.0 h at influent COD of 4000 mg L⁻¹; Specific methanogenic activity: 0.31 and 0.73 g CODCH₄ g⁻¹ VSS d⁻¹ for the first and second feed	[20]
Olive mill effluent	UMAR	5–48	~7	240–120	5–48	81–87	-	-	Maximum biogas production: 1.2 m3 m-3 d-1	[19]

Notes: ^{a} *UAF: upflow anaerobic filter; FBR: anaerobic fixed bed reactor; UASB: upflow anaerobic sludge blanket reactor; AF/BAF: integrated anaerobic/aerobic filter system; UASFB: hybrid upflow anaerobic sludge-filter bed reactor; JBlLAFB: full-scale jet biogas internal loop anaerobic fluidized bed. UMAR: upflow multistage anaerobic reactor; TCOD: total chemical oxygen demand; HRT: hydraulic retention time; OLR: organic loading rates; SCOD: soluble chemical oxygen demand; VFA: volatile fatty acids; VSS: volatile suspended solids; VS: volatile solids.*

6.4 SPECIFIC DESIGN AND OPERATIONAL CONSIDERATIONS

Within the high rate anaerobic treatment technologies, the immobilization of the microorganisms in the fixed bed reactors and the formation of the granules in UASB have been recognized as the two most frequently used anaerobic techniques to reduce the HRT during the anaerobic digestion of wastewater having low organic matter concentration [1,21]. The upflow fixed bed and UASB reactors are discussed in the subsequent sections.

6.4.1 UPFLOW FIXED BED REACTORS

The upflow fixed bed reactors or upflow anaerobic filter (UAF) is one of the earlier designs, and its characteristics are well defined [3,22,40]. The UAF is a relatively simple technology; and in engineering terms, it is not as complex as fluidized bed reactors, and in biological terms, it does not require the formation of a granular sludge, a prerequisite for the UASB reactor, which is usually very difficult to maintain. Also, fixed-film processes are inherently stable and resistant to organic and hydraulic shock loading conditions [6]. Since there is no provision made for intentional wastage of excess biomass from the filters similar to UASB reactors, clogging occurs with continued operation [1].

To accommodate the accumulation of non-attached biomass without plugging of the bed, the early designs of low voidage, rock-packed reactors have largely been replaced by systems that incorporate synthetic packing materials [6,41]. The microorganisms' immobilization on surfaces of synthetic carrier material is an increasingly used strategy to enhance biological treatment. The preferable features of the immobilized cells in comparison to the non-immobilized counterparts includes lower sensitivity to toxic loads, greater catalytic stability, longer microbial residence time, more tolerance to oligotrophic conditions and lower biomass washout risk [42,43].

Nikolaeva et al. [40] used waste tire rubber and zeolite as microorganism immobilization supports; Fia et al. [6] used blast furnace cinders, polyurethane foam and crushed stone as the supporting materials with porosities of 53%, 95% and 48%, respectively. Other materials that have been utilized for the adhesion of biomass included brick ballasts [44], Raschig rings (1.2 cm diameter ×1.2 cm) [45], low and high-density polyethylene media [3,22,35,41], polypropylene pall rings [46], non-woven disks [47], porous polyurethane foam [48] and modified porphyritic andesite (WRS) as ammonium adsorbent and bed material [49]. Rajagopal [1] suggested the unclogging procedures with fluidization by gas re-circulation or using liquid, which can be applied whenever clogging occurs. Such a problem can also be overcome by using UAF filled (80% or lower by volume) with low-density floating media [22]. Recently, biofilm reactors have also attracted more attention, especially for treatment of wastewaters containing bio-recalcitrant, inhibitory and toxic compounds [50,51]. However, there is a necessity of (i) post-treatment to reach the discharge standards for organic matter, nutrients (e.g., NH_4^+, PO_4^{3-}, S_2) and pathogens; and (ii) purification of biogas [22,52].

6.4.2 UASB REACTORS

The chief characteristics of a UASB reactor that makes it the established high-rate anaerobic digester worldwide (particularly in tropical countries) is the availability of granular or flocculent sludge, allowing it to achieve high organic matter (COD) removal efficiencies without the need of a support material [21,34,52,53]; the natural turbulence caused by the rising gas bubbles enhances the reactor content mixing and provides efficient wastewater and biomass contact. Therefore, mechanical mixing is not required, thus significantly reducing the energy demand and its associated cost.

Nevertheless, there are still unresolved issues in the anaerobic treatment technology. One of the major drawbacks of these reactors has been the requirement of a long solids retention time, which is not associated with the increasing volume of sludge produced from industrial and human activities [21]. Garcia et al. [34] listed various factors affecting granulation and, then, efficiency in UASB reactors viz. composition and concentration of organic matter in wastewater to be treated, operating temperature, pH, high ammonia nitrogen concentrations, presence of polyvalent cations, hydrodynamic conditions, inoculated seed and the production of exo-cellular polymeric substances by anaerobic bacteria. Rajagopal et al. [29] mentioned that with poor sedimentation characteristics in high loaded anaerobic reactors with suspended solids, the active biomass can be washed out from the reactor with the effluent, causing digester instability. Other drawbacks have been the long start-up period, flotation and disintegration of granular sludge, sludge bulking, deterioration of performance at low temperatures, high sulfate concentration, impure biogas and insufficient removal of organic matters, pathogens and nutrients in the final effluent, thereby failing to comply with the local standards for discharge or reuse [39,52,54,55].

6.4.3 INTEGRATED APPROACH: MODIFIED CONFIGURATIONS AND COMBINED SYSTEMS

The researchers have been trying to find out new technologies to enhance the performance of anaerobic digesters, especially on the effluent quality, start up and biogas purification, in order to develop a global sustainable wastewater treatment technology. Rajagopal et al. [29] proposed the hybrid upflow anaerobic sludge-filter bed (UASFB) reactors, in which the filter bed is located above the sludge bed zone, to reduce the risk of biomass washout from the reactor, especially at high loading rates. This configuration also improved the hydrodynamics characteristics by minimizing the clogging and short circuiting problems inside the reactors.

Lim and Fox [39] developed an anaerobic/aerobic filter (AF/BAF) system for the treatment of dairy wastewater, primarily to remove organic matter and nitrogen simultaneously. The influent was blended with recirculated effluent (100%–300%) to allow for pre-denitrification in the AF, followed by nitrification in the BAF. The average COD removal efficiency was 79.8%–86.8% in the AF, and the average total nitrogen removal efficiency was 50.5%–80.8% in the AF/BAF system. They have concluded that linear velocity was a critical parameter to determine sloughing of biomass in the AF.

Huang et al. [56] studied the mesophilic two-phase anaerobic system for the treatment of maize ethanol wastewater, particularly in terms of start-up, cultivation and the morphology of mature granular sludge in an improved methanogenic UASB reactor. By gradually increasing volumetric loading rate and regulating internal circulation, the UASB reactor developed bigger size granules (more than 1.3 mm), and hence, a quick start-up was achieved. They observed that the microfloras of mature methanogenic granular sludge were mainly Brevibacterium and filamentous bacteria.

A three-phase separator configuration increased the COD removal by 20%, biogas yield by 29% and decreased biomass wash out by 25% in an UASB treating fruit canning wastewater [57]. An intermittent feeding strategy at an OLR of 4.8 g COD L^{-1} d^{-1} of olive mill wastewater applied to an inverted anaerobic sludge blanket (IASB) reactor improved long chain fatty acids mineralization, prevented biomass washout and yielded 1.4 m^3 CH_4 m_{-3} d^{-1} [19]. Diez et al. [58] reported that an anaerobic membrane bioreactor (AnMBR) removed 97% of the COD of oil and greasy wastewater (COD of 22 g L^{-1}) at an OLR of 5.1 kg COD m^{-3} d^{-1}. Kim et al. [59] implemented a high-rate two-phase system (OLR = 6.5 g COD L^{-1} d^{-1}) of an anaerobic sequencing batch reactor (ASBR) and an upflow anaerobic sludge blanket (UASB) reactor in series to treat synthetic dairy wastewater treatment. The overall lipid and COD removals were about 80%.

A three-stage configuration of a pre-acidification tank and sequential upflow anaerobic sludge bed reactors (UASBRs) at OLRs of 2.8–7 g COD L^{-1} d^{-1} has been used to treat raw cheese whey with an effluent recycling achieved COD removal of 50%–92% and fat removal of 63%–89% [60]. An upflow anaerobic packed bed (UAPB) reactor filled with seashell treated cheese whey at OLRs of 1.6 to 9.9 g COD L^{-1} d^{-1}, and HRTs of 6–24 h at 25 °C removed 95% of COD [61].

In order to upgrade the quality of anaerobically treated effluent to a level recommended for irrigation, Yasar and Tabinda [62] integrated the UASB reactor with UV and AOPs (advanced oxidation processes) (Ozone, H_2O_2/UV, Fenton and photo-Fenton) primarily for complete color and COD removal and disinfection of pathogens.

6.5 MASS TRANSFER CONSIDERATIONS

Rajagopal [1] described that substrate mass transfer into biomass occurs most rapidly when the bio-particle surface-to-volume ratio is high. According to that research, the suspended growth anaerobic processes generate relatively small bio-particles with optimal surface-to-volume ratios. Fixed film or attached growth reactors are often considered to be susceptible to bio-film surface area limitations. In this respect, Chen et al. [63] concluded that the bed expansion ratio of a super-high rate anaerobic bioreactor correlated positively with superficial gas and liquid velocities, while maximum bed sludge content and maximum bed contact time between sludge and liquid correlated negatively. Ganesh et al. [3] addressed the effect of media design on treatment performance and stated that physical parameters, like the type of media, its size and shape, affect the performance of waste treatment. Chen et al. [64] modeled the dynamic behavior and concentration distribution of granular sludge in a super-high-rate spiral anaerobic bioreactor (SSAB) and found that these two parameters depend on the ecological environment of microbial communities and substrate degradation efficiency along the bed's height. The sludge transport efficiency of up-moving biogas in SSAB is less than that in a UASB reactor. Yang et al. [65] analyzed the mass transfer in tubular membrane anaerobic bioreactors operated under gas-lift two-phase flow using fluid dynamic modeling. They found that the results with water were in contrast to those with sludge. The sludge filterability strongly influences the transmembrane pressure, and there is a difference between the mass transfer capacity at the noses and the tails of the gas bubbles. Feng et al. [66] reported a rapid-mass transfer in a fluidized bed reactor using brick particles as carrier materials; at increasing OLRs from 7.37 to 18.52 kg COD m^{-3} d^{-1} and HRT of 8 h, COD removals between 65% and 75% was obtained. To minimize non-idealities in reactor hydraulics, most anaerobic reactor designs utilize proprietary systems for enhancing process mixing. Anaerobic contact systems rely on mechanical or gas recirculation systems that must be properly sized for the specific application [1].

6.6 ENERGY PRODUCTION ESTIMATION

The production of a useful and valuable product during agro-food waste-water treatment, such as hydrogen and/or methane, could help to lower treatment costs [1,67]. Using a single-chamber microbial electrolysis cell with a graphite-fiber brush anode, Wagner et al. [67] generated hydrogen gas at the rate of 0.9–1.0 m^3 m^{-3} day^{-1} using a full-strength or diluted swine wastewater. Under the best conditions, the specific hydrogen production rate of 270 mL H_2 g^{-1} MLVSS d^{-1} (or 3310 mL H_2 L^{-1} d^{-1}) and a hydrogen yield of 172 mL H_2 g^{-1} COD removed were obtained for the treatment of alcohol distillery wastewater containing high potassium and sulfate in an anaerobic sequencing batch reactor [68].

Nieto et al. [7] obtained methane yields ranging from 202 to 549 mL $CH_4 \cdot gVS_{fed}^{-1}$ at standard temperature and pressure (STP) from six wastes (beverage waste, milled apple waste, milk waste, yogurt waste, fats and oils (F&O) from dairy wastewater treatment, F&O and cattle manure). They reported that methane content in biogas ranged from 58% to 76%. Rajagopal et al. [69] developed a process combining anaerobic digestion and anoxic/oxic treatment to treat swine wastewater and obtained 5.9 Nm^3 of CH_4 m^{-3} of slurry added. In another study, Labatut et al. [70] observed that co-digestion of dairy manure with easily-degradable substrates increases the specific methane yields when compared to manure-only. For instance, co-digestion of dairy manure with used oil (ratio: 75:25 on volatile solids basis) produced more methane yield (361 Nm^3 $ton^{-1}VS_{added}$) than dairy manure with cheese whey (ratio: 75:25 on VS basis; methane yield: 252 Nm^3 $ton^{-1}VS_{added}$). Ogejo and Li [71] obtained a biogas yield ranging from 0.072 to 0.8 m^3 g^{-1} VS (methane content ranging from 56% to 70%) by co-digesting flushed dairy manure and turkey processing wastewater. The biogas produced was enough to run a 50 kW generator to produce electricity for about 5.5 and 9 h for the 1:1 and 1:2 feed mixtures. In addition to the electricity to be produced, other possible revenues, such as carbon credits, renewable energy credits, green tags for electricity, putting a value to the environmental benefits of anaerobic digestion or subsidies from grants or other incentives programs to make the system economically viable should be considered. On average, 18.5–40 kg of VS added in to the anaerobic digestion system can produce a biogas yield of 10 ± 5 m^3, when 65% VS removal is achieved [72]. This indicates a daily electricity generation of 12.5–33.6 kWh from biogas, on the assumption that the generator efficiency is 35%–50%. In addition, a daily heat energy profit of 17.8–46.5 kWh from biogas can be estimated. One cubic meter of biogas obtained while co-digesting dairy manure and animal fat is equivalent to 20 MJ of heat energy [73]. When used as fuel for a co-generator, 1 m^3 of biogas can produce 1.7 kWh of electricity and 7.7 MJ of heat. In addition to co-digestion, Esposito et al. [74] suggested several pre-treatment techniques that can be applied to increase further the biogas production, such as mechanical comminution, solid–liquid separation, bacterial hydrolysis and alkaline addition at high temperature, ensilage, alkaline, ultrasonic and thermal pre-treatments.

6.7 TOXICITY

Sodium toxicity is a common problem causing inhibition of anaerobic digestion. Digesters treating highly concentrated wastes, such as food and concentrated animal manure, are likely to suffer from partial or complete inhibition of methane-producing consortia, including methanogens [75,76]. Zhao et al. [10] confirmed that organofluorine compounds, such as 4-fluorophenol (p-FP), 4-fluorobenzoic acid (p-FB) and 4-fluoroaniline (p-FA), have a potential toxicity on methanogenesis and biodegradability. Procházka et al. [77] described that high ammonia nitrogen concentration (especially the unionized form) (4.0 g L^{-1}) would inhibit methane production, while low ammonia nitrogen concentration (0.5 g L^{-1}) could cause low methane yield, loss of biomass (as VSS) and loss of the acetoclastic methanogenic activity. Chen et al. [78] indicated that certain ions, such as Na$^+$, Ca^{2+} and Mg^{2+}, were found to be antagonistic to ammonia inhibition, a phenomenon in which the toxicity of one ion is decreased by the presence of other ion(s). At high concentrations, potassium, light metals ions (Na, K, Mg, Ca and Al) and other salts can also interrupt cell function [78].

6.7.1 TOXICITY CONTROL STRATEGIES

Vyrides et al. [79] indicated that glycine-betaine (GB), an organic compound, can cause antagonism against sodium toxicity. However, using GB to decrease sodium toxicity in commercial scale anaerobic digesters would be too costly [75]. Suwannoppadol et al. [75] described that when grass clippings were added at the onset of anaerobic digestion of acetate containing a sodium concentration of 7.8 g $Na^+ L^{-1}$, a total methane production of about 8 L CH_4 L^{-1} was obtained, whereas no methane was produced in the absence of grass leaves. Another way of tackling the sodium salts problem is by allowing the anaerobic sludge to acclimate to high sodium concentrations [80], but this technique requires time for the methanogens to adapt to the saline conditions, which in turn, results in a prolonged period before the anaerobic reactor can achieve its full-loading capacity. Zhao et al. [10] concluded that adsorption was the main removal mechanism for the three F-substituent aromatics, such as 4-fluorophenol (p-FP), 4-fluorobenzoic acid (p-FB) and 4-fluoroaniline (p-FA). To overcome ammonia toxicity, many strategies have been suggested: chemical precipitation, pH and temperature control [78]; use of carbon fiber textiles [81]; acclimation of methanogenic consortia to high ammonia levels [82]; and ammonia stripping and adjustment of the C:N ratio of feedstock [83]. Uludag-Demirer et al. [84] and Wang et al. [49] described the physical, chemical and biological methods, such as addition of ammonium-selective adsorbent, ammonium removal by forming struvite precipitation or a biological anoxic/oxic (A/O) process. Among these methods, ammonium removal by adding ammonium-selective adsorbents could be the most attractive and practical, because of its easier operation and economic impact, and the ammonium-saturated adsorbents can be further used as nitrogen fertilizer [49].

6.8 MODELING ADVANCES

Various kinetic equations reported for anaerobic processes [3,85] generally relied on Monod's equation. Monod's equation is based on the growth rate and the substrate utilization rate during biodegradation. Wong et al. [86] determined the biological kinetic constants for a laboratory anaerobic bench scale reactors (ABSR) treating palm oil mill effluent. The investigation showed that the growth yield (Y_G), specific biomass decay (b), maximum specific biomass growth rate (μ_{max}), saturation constant (K_s) and critical retention time (Θ_c) were in the range of 0.990 g VSS g^{-1} COD$_{removed}$ d^{-1}, 0.024 d^{-1}, 0.524 d^{-1}, 203.433 g COD L^{-1} and 1.908 d, respectively.

Fuzzato et al. [26] used simplified first-order kinetics for modeling a mesophilic anaerobic sequencing batch biofilm reactor (ASBBR) treating lipid-rich wastewater. Nikolaeva et al. [40] also observed that a first-order kinetic model described the experimental results obtained for the upflow fixed bed digesters treating dairy manure well. In addition, they also concluded that the first-order model was adequate for assessing the effect of HRT on the removal efficiency and methane production. Ganesh et al. [3] used a modified Stover-Kincannon kinetic model to predict the performance of anaerobic fixed bed reactors treating winery wastewater. In a similar study, Rajagopal et al. [87] applied bio-kinetic models, such as the modified Stover-Kincannon and second-order kinetic models for the upflow anaerobic filters treating high strength fruit canning and cheese dairy wastewaters. Fia et al. [6] also used the modified Stover-Kincannon and second-order kinetic models for the experimental data obtained from the upflow anaerobic fixed bed reactors treating coffee bean processing wastewater. Abdurahman et al. [85] employed kinetic equations from Monod, Contois and Chen and Hashimoto to describe the kinetics of palm oil mill effluent (POME) treatment in a membrane anaerobic system (MAS) at organic loading rates ranging from 1 to 11 kg COD m^{-3} d^{-1}. Kaewsuk et al. [88] conducted a pilot scale experiment to investigate the performance of the membrane sequencing batch reactor (MSBR) treating dairy wastewater and checked the suitability of the kinetics for an engineering design. The kinetic coefficients K_s, k, k_d, Y and mm were found to be 174-mg COD L^{-1}, 7.42 d^{-1}, 0.1383 d^{-1}, 0.2281 d^{-1} and 1.69 d^{-1}, respectively.

Recently, there has been a move by the International Water Association's (IWA) Task Group for Mathematical Modeling of Anaerobic Digestion Processes to develop a common model called Anaerobic Digestion Model No. 1 (ADM1) that can be used by researchers and practitioners [89,90]. Lee et al. [91] examined the application of the ADM1 for mathematical modeling of anaerobic process using a lab-scale temperature-phased anaerobic digestion (TPAD) process. Sensitivity analysis showed that $k_{m.process}$ (maximum specific uptake rate) and $K_{S.process}$ (half saturation value) had high sensitivities to model components. They have concluded that simulation with estimated parameters showed good agreement with experimental results in the case of methane production, uptake of acetate, soluble (SCOD) and total chemical oxygen demand (TCOD). The structure and properties of a microbial community may be influenced by process operation and, in their turn, also determine the reactor functioning. In order to adequately describe these phenomena, Ramirez et al. [89] emphasized that mathematical models need to consider the underlying microbial diversity. In order to demonstrate this contribution, they have extended the ADM1 to describe microbial diversity between organisms of the same functional group. Boubaker and Ridha [92] used the ADM1 model to simulate the mesophilic anaerobic co-digestion of olive mill wastewater (OMW) with olive mill solid waste (OMSW). The results indicated that the ADM1 model could simulate with good accuracy: gas flows, methane and carbon-dioxide contents, pH and total volatile fatty acids (TVFA) concentrations of effluents for various feed concentrations digested at different HRTs and especially at HRTs of 36 and 24 days. Furthermore, effluent alkalinity and ammonium nitrogen were also successfully predicted.

6.9 TECHNOLOGY ASSESSMENTS

Onsite industrial wastewater anaerobic treatment requires systems with a significant capital cost and incur increasing expenses for successful long-term operation, control and maintenance. Farhan [93] evaluated a high-rate digestion system for brewery wastewater technically and economically. The technical evaluating criteria consist of, among commonly used engineering design criteria, such as hydraulic and organic loading rates, wastewater characteristics and layout and space requirements, factors that reflect the dynamics of technology development. The abilities of the anaerobic high rate bioreactors to meet the regulatory requirements reliably with flexibility for future upgrading are some of the technical evaluating factors [93]. The energy savings and renewable energy credits are among the economical assessment criteria [93].

Gebrezgabher et al. [94] analyzed the economic performance of anaerobic digestion of a biogas plant using the net present value (NPV) and internal rate of return (IRR). They conclude that the uncertainty of the increasingly tightened regulations regarding the effluent of anaerobic treatment, the quality and value of the digestate and the high investment and operating costs limit the on-farm applications of anaerobic digestion of agro wastes.

6.10 SUMMARY AND CONCLUSIONS

A critical analysis of the literature reveals that there is a strong possibility and need to enhance the performance of high rate anaerobic biogas reactors. This technique has many advantages over other conventional methods. However, the challenges associated with the digester operation at lower HRT and higher OLR need to be addressed, which include biomass washout, clogging, short-circuiting, process inhibitions, poor final effluent and biogas quality. Different materials (polyethylene, polypropylene pall rings, polyurethane foam, carbon felt and waste tire rubber) had been tried as packing material in the anaerobic fixed bed reactors, depending upon their availability and other specifications, such as material properties, cost, etc. These packing materials would help to reduce hydraulic retention time, which in turn lessens the required volume of the reactor, and ultimately, the cost could be reduced. The practical aspects of using pure microbial film as magnifying microbial layers should be looked into. In the case of UASB reactors, the following important aspects should be looked into: (i) enhancing the start-up and granulation in UASB reactors; (ii) coupling with post-treatment unit to overcome the temperature constraint; and (iii) improving the removal efficiencies of the organic matter, nutrients and pathogens in the final effluent.

When talking about toxicity, there are many soluble organic and inorganic materials that can be either stimulatory or inhibitory. A good example of this is the effect of ammonia nitrogen on the anaerobic digestion process. When potential inhibitory materials are slowly increased within the environment, many biological organisms can rearrange their metabolic resources, thus overcoming the metabolic block produced by the normally inhibitory material. Under shock load conditions, sufficient time is not available for this rearrangement to take place. Finally, there is the possibility of antagonism and synergism effects of using different organic wastes as co-substrates. Antagonism is defined as a reduction of the toxic effect of one substance by the presence of another. Synergism is defined as an increase in the toxic effect of one substance by the presence of another. This is an important consideration when designing for potential cation toxicity. Additional research efforts are essential to get more insight about the stable performance of the digesters against various process inhibitors, such as ammonia, sodium, sulfur, etc. While lessening the economic losses, vigilant substrate management and early detection of inhibitions are critical.

REFERENCES

1. Rajagopal, R. Treatment of Agro-Food Industrial Wastewaters Using UAF and Hybrid UASB-UAF Reactors. Ph.D. Thesis, Indian Institute of Technology Roorkee, Roorkee, India, 2008.
2. Badshah, M.; Parawira, W.; Mattiasson, B. Anaerobic treatment of methanol condensate from pulp mill compared with anaerobic treatment of methanol using mesophilic UASB reactors. Bioresour. Technol. 2012, 125, 318–327.
3. Ganesh, R.; Rajagopal, R.; Torrijos, M.; Thanikal, J.M.; Ramanujam, R. Anaerobic treatment of winery wastewater in fixed bed reactors. Bioprocess Biosyst. Eng. 2010, 33, 619–628.
4. Meksi, N.; Haddar, W.; Hammami, S.; Mhenni, M.F. Olive mill wastewater: A potential source of natural dyes for textile dyeing. Ind. Crops Prod. 2012, 40, 103–109.
5. MPOC (Malaysian Palm Oil Council). Available online: http://www.mpoc.org.my (accessed on 5 March 2013).
6. Fia, R.L.; Matos, A.T.; Borges, A.C.; Fia, R.; Cecon, P.R. Treatment of wastewater from coffee bean processing in anaerobic fixed bed reactors with different support materials: Performance and kinetic modeling. J. Environ. Manag. 2012, 108, 14–21.
7. Nieto, P.P.; Hidalgo, D.; Irusta, R.; Kraut, D. Biochemical methane potential (BMP) of agro-food wastes from the Cider Region (Spain). Water Sci. Technol. 2012, 66, 1842–1848.
8. Veronafiere (Ente Autonomo per le Fiere di Verona). Vinitaly analysis. Worldwide Wine: The Sector Scenario, Production, Consumption and Trade on a World Scale under the Magnifying Glass at Vinitaly. Available online: http://www.vinitaly.com/pdf/cartellaStampa/5gbCsVinitaly12_SituazioneMondoItalia_23marzo.pdf (accessed on 5 March 2013).
9. Un, U.T.; Altay, U.; Koparal, A.S.; Ogutveren, U.B. Complete treatment of olive mill wastewaters by electrooxidation. Chem. Eng. J. 2008, 139, 445–452.

10. Zhao, Z.-Q.; Xu, L.-L.; Li, W.-B.; Wang, M.-Z.; Shen, X.-L.; Mae, G.-S.; Shena, D.-S. Toxicity of three F-substituent aromatics in anaerobic systems. J. Chem. Technol. Biotechnol. 2012, 87, 1489–1496.

11. Wei, C.; Zhang, T.; Feng, C.; Wu, H.; Deng, Z.; Wu, C.; Lu, B. Treatment of food processing wastewater in a full-scale jet biogas internal loop anaerobic fluidized bed reactor. Biodegradation 2011, 22, 347–357.

12. Rupani, P.F.; Singh, R.P.; Ibrahim, M.H.; Esa, N. Review of current palm oil mill effluent (POME) treatment methods: Vermicomposting as a sustainable practice. World Appl. Sci. J. 2010, 10, 1190–1201.

13. Alkaya, E.; Demirer, G.N. Anaerobic acidification of sugar-beet processing wastes: Effect of operational parameters. Biomass Bioenergy 2011, 35, 32–39.

14. Gotmare, M.; Dhoble, R.M.; Pittule, A.P. Biomethanation of dairy waste water through UASB at mesophilic temperature range. Int. J. Adv. Eng. Sci. Technol. 2011, 8, 1–9.

15. Ersahin, M.E.; Ozgun, H.; Dereli, R.K.; Ozturk, I. Anaerobic treatment of industrial effluents: An overview of applications. Waste Water—Treatment and Reutilization; InTech: New York, NY, USA, 2011. Available online: http://www.intechopen.com/books/waste-water-treatment-and-reutilization/anaerobic-treatment-of-industrial-effluents-an-overview-of-applications (accessed on 5 March 2013).

16. Senturk, E.; Ince, M.; Engin, O.G. Treatment efficiency and VFA composition of a thermophilic anaerobic contact reactor treating food industry wastewater. J. Hazard. Mater. 2010, 176, 843–848.

17. Ersahin, M.E.; Dereli, R.K.; Ozgun, H.; Donmez, B.G.; Koyuncu, I.; Altinbas, M.; Ozturk, I. Source based characterization and pollution profile of a baker's yeast industry. Clean-Soil Air Water 2011, 39, 543–548.

18. Passeggi, M.; Lopez, I.; Borzacconi, L. Integrated anaerobic treatment of dairy industrial wastewater and sludge. Water Sci. Technol. 2009, 59, 501–506.

19. Gonçalves, M.R.; Costa, J.C.; Marques, I.P.; Alves, M.M. Strategies for lipids and phenolics degradation in the anaerobic treatment of olive mill wastewater. Water Res. 2012, 46, 1684–1692.

20. Sun, L.; Wan, S.; Yu, Z.; Wang, Y.; Wang, S. Anaerobic biological treatment of high strength cassava starch wastewater in a new type up-flow multistage anaerobic reactor. Bioresour. Technol. 2012, 104, 280–288.

21. Chong, S.; Sen, T.K.; Kayaalp, A.; Ang, H.M. The performance enhancements of upflow anaerobic sludge blanket (UASB) reactors for domestic sludge treatment—A State-of-the-art review. Water Res. 2012, 46, 3434–3470.

22. Rajagopal, R.; Ganesh, R.; Escudie, R.; Mehrotra, I.; Kumar, P.; Thanikal, J.V.; Torrijos, M. High rate anaerobic filters with floating supports for the treatment of effluents from small-scale agro-food industries. Desalin. Water Treat. 2009, 4, 183–190.

23. Esparza, S.M.; Solís, M.C.; Herná, J.J. Anaerobic treatment of a medium strength industrial wastewater at low-temperature and short hydraulic retention time: A pilot-scale experience. Water Sci. Technol. 2011, 64, 1629–1635.

24. Shastry, S.; Nandy, T.; Wate, S.R.; Kaul, S.N. Hydrogenated vegetable oil industry wastewater treatment using UASB reactor system with recourse to energy recovery. Water Air Soil Pollut. 2010, 208, 323–333.

25. Won, S.G.; Lau, A.K. Effects of key operational parameters on biohydrogen production via anaerobic fermentation in a sequencing batch reactor. Bioresour. Technol. 2011, 102, 6876–6883.

26. Fuzzato, M.C.; Tallarico Adorno, M.A.; de Pinho, S.C.; Ribeiro, R.; Tommaso, G. Simplified mathematical model for an anaerobic sequencing batch biofilm reactor treating lipid-rich wastewater subject to rising organic loading rates. Environ. Eng. Sci. 2009, 26, 1197–1206.

27. Shanmugam, A.S.; Akunna, J.C. Comparing the performance of UASB and GRAB-BR treating low strength wastewaters. Water Sci. Technol. 2008, 58, 225–232.

28. Bialek, K.; Kumar, A.; Mahony, T.; Lens, P.N.L.; Flaherty, V.O. Microbial community structure and dynamics in anaerobic fluidized-bed and granular sludge-bed reactors: Influence of operational temperature and reactor configuration. Microb. Biotechnol. 2012, 5, 738–775.

29. Rajagopal, R.; Mehrotra, I.; Kumar, P.; Torrijos, M. Evaluation of a hybrid upflow anaerobic sludge-filter bed reactor: Effect of the proportion of packing medium on performance. Water Sci. Technol. 2010, 61, 1441–1450.

30. Gnansounou, E. Production and use of lignocellulosic bioethanol in Europe: Current situation and perspectives. Bioresour. Technol. 2010, 101, 4842–4850.

31. Nkemka, V.N.; Murto, M. Biogas production from wheat straw in batch and UASB reactors: The roles of pretreatment and seaweed hydrolysate as a co-substrate. Bioresour. Technol. 2013, 128, 164–172.

32. Alvira, P.; Tomás-Pejó, E.; Ballesteros, M.; Negro, M.J. Pretreatment technologies for an efficient bioethanol production process based on enzymatic hydrolysis: A review. Bioresour. Technol. 2010, 101, 4851–4861.

33. Bosco, F.; Chiampo, F. Production of polyhydroxyalcanoates (PHAs) using milk whey and dairy wastewater activated sludge Production of bioplastics using dairy residues. J. Biosci. Bioeng. 2010, 109, 418–421.

34. Garcia, H.; Rico, C.; Garcia, P.A.; Rico, J.L. Flocculants effect in biomass retention in a UASB reactor treating dairy manure. Bioresour. Technol. 2008, 99, 6028–6036.

35. Yu, Y.-C.; Gao, D.-W.; Tao, Y. Anammox start-up in sequencing batch biofilm reactors using different inoculating sludge. Appl. Microbiol. Biotechnol. 2012.

36. Zhang, T.; Yan, Q.M.; Ye, L. Autotrophic biological nitrogen removal from saline wastewater under low DO. J. Chem. Technol. Biot. 2010, 85, 1340–1345.

37. Hendrickx, T.L.G.; Wang, Y.; Kampman, C.; Zeeman, G.; Temmink, H.; Buisman, C.J.N. Autotrophic nitrogen removal from low strength waste water at low temperature. Water Res. 2012, 46, 2187–2193.

38. Dosta, J.; Fernandez, I.; Vazquez-Padin, J.R.; Mosquera-Corral, A.; Campos, J.L.; Mata-Alvarez, J.; Mendez, R. Short- and long-term effects of temperature on the anammox process. J. Hazard. Mater. 2008, 154, 688–693.

39. Lim, S.J.; Fox, P. A kinetic analysis and experimental validation of an integrated system of anaerobic filter and biological aerated filter. Bioresour. Technol. 2011, 102, 10371–10376.

40. Nikolaeva, S.; Sanchez, E.; Borja, R.; Raposo, F.; Colmenarejo, M.F.; Montalvo, S.; Jiménez-Rodríguez, A.M. Kinetics of anaerobic degradation of screened dairy manure by upflow fixed bed digesters: Effect of natural zeolite addition. J. Environ. Sci. Health Part A Toxic/Hazard. Substan. Environ. Eng. 2009, 44, 146–154.

41. Koupaie, E.H.; Moghaddam, M.R.A.; Hashemi, S.H. Evaluation of integrated anaerobic/aerobic fixed-bed sequencing batch biofilm reactor for decolorization and biodegradation of azo dye Acid Red 18: Comparison of using two types of packing media. Bioresour. Technol. 2013, 127, 415–421.

42. Gómez-De Jesús, A.; Romano-Baez, F.J.; Leyva-Amezcua, L.; Juárez-Ramírez, C.; Ruiz-Ordaz, N.; Galíndez-Mayer, J. Biodegradation of 2,4,6-richlorophenol in a packed-bed biofilm reactor equipped with an internal net draft tube riser for aeration and liquid circulation. J. Hazard. Mater. 2009, 161, 1140–1149.

43. González, A.J.; Gallego, A.; Gemini, V.L.; Papalia, M.; Radice, M.; Gutkind, G.; Planes, E.; Korol, S.E. Degradation and detoxification of the herbicide 2,4-dichlorophenoxyacetic acid (2,4-D) by an indigenous Delftia sp. strain in batch and continuous systems. Int. Biodeter. Biodegr. 2012, 66, 8–13.

44. Satyawali, Y.; Pant, D.; Singh, A.; Srivastava, R.K. Treatment of rayon grade pulp drain effluent by upflow anaerobic fixed packed bed reactor (UAFPBR). J. Environ. Biol. 2009, 30, 667–672.

45. Ahn, J.-H. Nitrogen requirement for the mesophilic and thermophilic upflow anaerobic filters of a simulated paper mill wastewater. Korean J. Chem. Eng. 2008, 25, 1022–1025.

46. Deshannavar, U.B.; Basavaraj, R.K.; Naik, N.M. High rate digestion of dairy industry effluent by upflow anaerobic fixed-bed reactor. J. Chem. Pharma. Res. 2012, 4, 2895–2899.

47. Gao, F.; Zhang, H.; Yang, F.; Qiang, H.; Zhang, G. The contrast study of anammoxdenitrifying system in two non-woven fixed-bed bioreactors (NFBR) treating different low C/N ratio sewage. Bioresour. Technol. 2012, 114, 54–61.

48. Ji, G.; Wu, Y.; Wang, C. Analysis of microbial characterization in an upflow anaerobic sludge bed/biological aerated filter system for treating microcrystalline cellulose wastewater. Bioresour. Technol. 2012, 120, 60–69.

49. Wang, Q.; Yang, Y.; Li, D.; Feng, C.; Zhang, Z. Treatment of ammonium-rich swine waste in modified porphyritic andesite fixed-bed anaerobic bioreactor. Bioresour. Technol. 2012, 111, 70–75.

50. Bajaj, M.; Gallert, C.; Winter, J. Biodegradation of high phenol containing synthetic wastewater by an aerobic fixed bed reactor. Bioresour. Technol. 2008, 99, 8376–8381.

51. Farhadian, M.; Duchez, D.; Vachelard, C.D.; Larroche, C. Monoaromatics removal from polluted water through bioreactors—A review. Water Res. 2008, 42, 1325–1341.

52. Mahmoud, N. High strength sewage treatment in a UASB reactor and an integrated UASB-digester system. Bioresour. Technol. 2008, 99, 7531–7538.

53. Elangovan, C.; Sekar, A.S.S. Application of Upflow anaerobic sludge blanket (UASB) reactor process for the treatment of dairy wastewater—A review. Nat. Environ. Pollut. Technol. 2012, 11, 409–414.

54. Lew, B.; Lustig, I.; Beliavski, M.; Tarre, S.; Green, M. An integrated UASB-sludge digester system for raw domestic wastewater treatment in temperate climates. Bioresour. Technol. 2011, 102, 4921–4924.

55. Li, J.; Hu, B.; Zheng, P.; Qaisar, M.; Mei, L. Filamentous granular sludge bulking in a laboratory scale UASB reactor. Bioresour. Technol. 2008, 99, 3431–3438.

56. Huang, J.P.; Liu, L.; Shao, Y.M.; Song, H.J.; Wu, L.C.; Xiao, L. Study on cultivation and morphology of granular sludge in improved methanogenic UASB. Appl. Mechan. Mater. 2012, 209–211, 1152–1157.

57. Wongnoi, R.; Songkasiri, W.; Phalakornkule, C. Influence of a three-phase separator configuration on the performance of an upflow anaerobic sludge bed reactor treating wastewater from a fruit-canning factory. Water Environ. Res. 2007, 79, 199–207.

58. Diez, V.; Ramos, C.; Cabezas, J.L. Treating wastewater with high oil and grease content using an anaerobic membrane bioreactor (AnMBR). Filtration and cleaning assays. Water Sci. Technol. 2012, 65, 1847–1853.

59. Kim, S.-H.; Shin, H.-S. Enhanced lipid degradation in an upflow anaerobic sludge blanket reactor by integration with an acidogenic reactor. Water Environ. Res. 2010, 82, 267–272.

60. Erdirencelebi, D. Treatment of high-fat-containing dairy wastewater in a sequential UASBR system: Influence of recycle. J. Chem. Technol. Biotechnol. 2011, 86, 525–533.

61. Najafpour, G.D.; Komeili, M.; Tajallipour, M.; Asadi, M. Bioconversion of cheese whey to methane in an upflow anaerobic packed bed bioreactor. Chem. Biochem. Eng. Q. 2010, 24, 111–117.

62. Yasar, A.; Tabinda, A.B. Anaerobic treatment of industrial wastewater by UASB reactor integrated with chemical oxidation processes: An overview. Pol. J. Environ. Stud. 2010, 19, 1051–1061.

63. Chen, X.-G.; Zheng, P.; Cai, J.; Qaisar, M. Bed expansion behavior and sensitivity analysis for super-high-rate anaerobic bioreactor. J. Zhejiang Univ. Sci. B. 2010, 11, 79–86.

64. Chen, X.G.; Zheng, P.; Qaisar, M.; Tang, C.J. Dynamic behavior and concentration distribution of granular sludge in a super-high-rate spiral anaerobic bioreactor. Bioresour. Technol. 2012, 111, 134–140.

65. Yang, J.; Vedantam, S.; Spanjers, H.; Nopens, I.; van Lier, J.B. Analysis of mass transfer characteristics in a tubular membrane using CFD modeling. Water Res. 2012, 46, 4705–4712.

66. Feng, Y.; Lu, B.; Jiang, Y.; Chen, Y.; Shen, S. Anaerobic degradation of purified terephthalic acid wastewater using a novel, rapid mass-transfer circulating fluidized bed. Water Sci. Technol. 2012, 65, 1988–1993.

67. Wagner, R.C.; Regan, J.M.; Oh, S.E.; Zuo, Y.; Logan, B.E. Hydrogen and methane production from swine wastewater using microbial electrolysis cells. Water Res. 2009, 43, 1480–1488.

68. Searmsirimongkol, P.; Rangsunvigit, P.; Leethochawalit, M.; Chavadej, S. Hydrogen production from alcohol distillery wastewater containing high potassium and sulfate using an anaerobic sequencing batch reactor. Int. J. Hydrog. Energ. 2011, 36, 12810–12821.

69. Rajagopal, R.; Rousseau, P.; Bernet, N.; Béline, F. Combined anaerobic and activated sludge anoxic/oxic for piggery wastewater treatment. Bioresour. Technol. 2011, 102, 2185–2192.

70. Labatut, R.A.; Angenent, L.T.; Scott, N.R. Biochemical methane potential and biodegradability of complex organic substrates. Bioresour. Technol. 2011, 102, 2255–2264.

71. Ogejo, J.A.; Li, L. Enhancing biomethane production from flush dairy manure with turkey processing wastewater. Appl. Energ. 2010, 87, 3171–3177.

72. Rajagopal, R.; Lim, J.W.; Mao, Y.; Chen, C.L.; Wang, J.Y. Anaerobic co-digestion of source segregated brown water (feces-without-urine) and food waste: For Singapore context. Sci. Total Environ. 2013, 443, 877–886.

73. Alberta Agriculture and Rural Development. Economic Feasibility of Anaerobic Digesters; Department of Agriculture and Rural Development: Edmonton, AB, Canada, 2008. Available online: http://www1.agric.gov.ab.ca/$department/deptdocs.nsf/all/agdex12280 (accessed on 5 March 2013).

74. Esposito, G.; Frunzo, L.; Giordano, A.; Liotta, F.; Panico, A.; Pirozzi, F. Anaerobic co-digestion of organic wastes. Rev. Environ. Sci. Biotechnol. 2012, 11, 325–341.

75. Suwannoppadol, S.; Ho, G.; Cord-Ruwisch, R. Overcoming sodium toxicity by utilizing grass leaves as co-substrate during the start-up of batch thermophilic anaerobic digestion. Bioresour. Technol. 2012, 125, 188–192.

76. Hierholtzer, A.; Akunna, J.C. Modelling sodium inhibition on the anaerobic digestion process. Water Sci. Technol. 2012, 66, 1565–1573.

77. Procházka, J.; Dolejš, P.; Máca, J.; Dohányos, M. Stability and inhibition of anaerobic processes caused by insufficiency or excess of ammonia nitrogen. Appl. Microbiol. Biotechnol. 2012, 93, 439–447.

78. Chen, Y.; Cheng, J.J.; Creamer, K.S. Inhibition of anaerobic digestion process: A review. Bioresour. Technol. 2008, 99, 4044–4064.

79. Vyrides, I.; Santos, H.; Mingote, A.; Ray, M.J.; Stuckey, D.C. Are compatible solutes compatible with biological treatment of saline wastewater? Batch and continuous studies using submerged anaerobic membrane bioreactors (SAMBRs). Environ. Sci. Technol. 2010, 44, 7437–7442.

80. Vyrides, I.; Stuckey, D.C. Adaptation of anaerobic biomass to saline conditions: Role of compatible solutes and extracellular polysaccharides. Enzym. Microb. Technol. 2009, 44, 46–51.

81. Sasaki, K.; Morita, M.; Hirano, S.; Ohmura, N.; Igarashi, Y. Decreasing ammonia inhibition in thermophilic methanogenic bioreactors using carbon fiber textiles. Appl. Microbiol. Biotechnol. 2011, 90, 1555–1561.

82. Abouelenien, F.; Nakashimada, Y.; Nishio, N. Dry mesophilic fermentation of chicken manure for production of methane by repeated batch culture. J. Biosci. Bioeng. 2009, 107, 293–295.

83. Resch, C.; Wörl, A.; Waltenberger, R.; Braun, R.; Kirchmayr, R. Enhancement options for the utilisation of nitrogen rich animal by-products in anaerobic digestion. Bioresour. Technol. 2011, 102, 2503–2510.

84. Uludag-Demirer, S.; Demirer, G.N.; Frear, C.; Chen, S. Anaerobic digestion of dairy manure with enhanced ammonia removal. J. Environ. Manag. 2008, 86, 193–200.

85. Abdurahman, N.H.; Rosli, Y.M.; Azhari, N.H. Development of a membrane anaerobic system (MAS) for palm oil mill effluent (POME) treatment. Desalination 2011, 266, 208–212.

86. Wong, Y.S.; Kadir, M.O.A.B.; Teng, T.T. Biological kinetics evaluation of anaerobic stabilization pond treatment of palm oil mill effluent. Bioresour. Technol. 2009, 100, 4969–4975.

87. Rajagopal, R.; Torrijos, M.; Kumar, P.; Mehrotra, I. Substrate removal kinetics in high-rate upflow anaerobic filters packed with low-density polyethylene media treating high-strength agro-food wastewaters. J. Environ. Manag. 2013, 116, 101–106.

88. Kaewsuk, J.; Thorasampan, W.; Thanuttamavong, M.; Seo, J.T. Kinetic development and evaluation of membrane sequencing batch reactor (MSBR) with mixed cultures photosynthetic bacteria for dairy wastewater treatment. J. Environ. Manag. 2010, 91, 1161–1168.

89. Ramirez, I.; Volcke, E.I.P.; Rajagopal, R.; Steyer, J.P. Application of ADM1 towards modelling biodiversity in anaerobic digestion. Water Res. 2009, 43, 2787–2800.

90. Girault, R.; Bridoux, G.; Nauleau, F.; Poullain, C.; Buffet, J.; Steyer, J.-P.; Sadowski, A.G.; Béline, F. A waste characterisation procedure for ADM1 implementation based on degradation kinetics. Water Res. 2012, 46, 4099–4110.

91. Lee, M.-Y.; Suh, C.-W.; Ahn, Y.-T.; Shin, H.-S. Variation of ADM1 by using temperature-phased anaerobic digestion (TPAD) operation. Bioresour. Technol. 2009, 100, 2816–2822.

92. Boubaker, F.; Ridha, B.C. Modelling of the mesophilic anaerobic co-digestion of olive mill wastewater with olive mill solid waste using anaerobic digestion model No. 1 (ADM1). Bioresour. Technol. 2008, 99, 6565–6577.

93. Farhan, M.H. High Rate Anaerobic Digester Systems for Brewery Wastewater Treatment and Electricity Generation: Engineering Design Factors and Cost Benefit Analysis. In Proceedings of The World Brewing Congress, Oregon Convention Centre, Portland, OR, USA, 28 July–1 August 2012; Available online: http://www.worldbrewingcongress.org/2012/Abstracts/AbstractsDetail.cfm?AbstractID=318(accessed on 5 March 2013).

94. Gebrezgabher, S.A.; Meuwissen, M.P.M.; Prins, B.A.M.; Lansink, A.G.J.M.O. Economic analysis of anaerobic digestion—A case of green power biogas plant in The Netherlands. NJAS Wagening. J. Life Sci. 2010, 57, 109–115.

PART III

PHARMACEUTICAL INDUSTRY

CHAPTER 7

Intensification of Sequencing Batch Reactors by Cometabolism and Bioaugmentation with *Pseudomonas putida* for the Biodegradation of 4-Chlorophenol

VICTOR M. MONSALVO, MONTSERRAT TOBAJAS,
ANGEL F. MOHEDANO, AND JUAN J. RODRIGUEZ

7.1 INTRODUCTION

Chlorophenols constitute an important group of hazardous compounds widely used in pesticides, herbicides and dyes manufacture. Among the monochlorophenols, 4-chlorophenol (4-CP) has been reported to be the less biodegradable [1] and more toxic compound (EC_{50} = 1.9 mg L^{-1}) [2] under aerobic conditions, and it has been found refractory in anaerobic systems. [3] 4-CP has been widely used as target toxic compound to evaluate the treatability of hardly biodegradable compounds by different biological systems such as conventional activated sludge, [4, 5] fluidized bed reactors, [6] packed-bed reactor, [7, 8] SBR, [9–11] rotating tubes biofilm

Printed with permission from Monsalvo VM, Tobajas M, Mohedano AF, and Rodriguez JJ. Intensification of Sequencing Batch Reactors by Cometabolism and Bioaugmentation with Pseudomonas putida *for the Biodegradation of 4-Chlorophenol. Journal of Chemical Technology and Biotechnology 87,9 (2012), DOI: 10.1002/jctb.3777.*

reactors, [12] rotating brush biofilm reactors, [13] moving bed biofilm reactors, [14] and membrane sequencing batch reactor. [15, 16]

Biological treatment of wastewater containing chlorophenols by conventional activated sludge (AS) often fails to achieve high removal efficiencies due to microbial inhibition [17] by interfering with energy transduction of cells. [18] Therefore, several strategies could be used to enhance the removal efficiency of hardly-biodegradable compounds. These methods include adaptation of the microbial population to the pollutant [19, 20] and/or the addition of alternative carbon sources such as phenol, acetate or sugars. These cosubstrates can act as inducing agents of oxidizing enzymes or by providing reducing capability for the degradation of recalcitrant compounds under aerobic conditions. Reports in the literature have concluded that cometabolism is a powerful way of enhancing chlorophenols biodegradation, when dealing with higher concentrations of those compounds. [9, 21] Several authors have shown that phenol is a better growth cosubstrate than biogenic compounds because of its similitude to chlorophenols. [22]

The introduction of specialist degrading bacteria in biological systems is a promising cost-effective intensification technique for improving both the removal efficiency and the start-up of bioreactors treating xenobiotics-bearing wastewaters. [7] The so-called bioaugmentation has been mostly applied for the remediation of soils contaminated with hazardous pollutants and also for treating wastewaters bearing hardly biodegradable compounds in small-scale AS systems. [23] However, its application to other biological systems has scarcely been studied so far. Nevertheless, 4-CP can be partially or completely removed by aerobic strains *Azotobacter, Alcaligenes, Rhodococcus, Phragmites* and specially *Pseudomonas* the most studied genus. The bioaugmentation of aerobic AS with *P. putida* has been successfully applied to improve the biological removal of chlorophenols batchwise. [24] It is noteworthy that the simple addition of allochthonous bacteria possessing metabolic capabilities does not guarantee enhanced transformation in a mixed culture. [25] So far, very few examples of successful bioaugmentation of AS units using either naturally isolated or genetically engineered microorganisms have been reported. [26]

Among the different biological systems used for the treatment of xenobiotics-bearing wastewaters, sequencing batch reactors (SBR) have received increasing attention due to some advantages of this technology such as easier control of the process, high nutrients removal efficiency and low energy requirements. In addition the operating conditions can be periodically modified to allow control of the abundance and activity of allochthonous strains in multi-species microbial communities [27] and to facilitate the survival and growth of the bacteria introduced. [28, 29] In this work, the effect of using phenol as cosubstrate and bioaugmentation of AS with *P. putida* for the biodegradation of 4-CP from synthetic wastewater by SBR was investigated.

7.2 EXPERIMENTAL

7.2.1 SBR DESCRIPTION AND OPERATION

The experiments were carried out in three SBRs of 2.1 L inoculated with AS (control SBR), *P. putida* or bioaugmented AS with *P. putida* (0.3%, w:w). The runs were conducted in a series of sequential stages of 12 h consisting of anoxic filling (1 h), aerated reaction (9.5 h), settling (1 h) and discharging (0.5 h). The reactors were operated at each 4-CP concentration investigated for at least 3 weeks, in order to ensure steady performance before collection of the corresponding data. The operating conditions were 30 ± 1 °C, neutral pH, 120 rpm, 5% volume exchange ratio, and 10.5 days hydraulic residence time (HRT). [9]

7.2.2 MICROORGANISMS AND GROWTH CONDITIONS

P. putida strain CECT 4064 was purchased from the Spanish Type Culture Collection (Valencia). Stock cultures were maintained at–40 °C in a nutrient medium supplemented with 15% (v/v) glycerol. *P. putida* was transferred to a nutrient medium containing 1 g beef extract, 2 g yeast extract, 5 g peptone and 5 g NaCl per litre of deionized water. The cell suspension resulting from the late exponential growth phase was subcultured in a mineral salts medium [30] with phenol (25 mg L^{-1}) as sole carbon source and stirred in a thermostated orbital shaker at 120 rpm, 30 °C and neutral pH for 10–12 h. The resulting culture was inoculated into the corresponding SBR.

The AS used as inoculum was collected from a cosmetics wastewater treatment plant. Biomass concentration in the reactors was maintained at 2500 ± 200 mg VSS L^{-1}. Organic load rates of 2.4 and 3.8 mg g^{-1} VSS day^{-1} of 4-CP and phenol, respectively, were used during the acclimation period, which lasted for 15 weeks.

7.2.3 WASTEWATER COMPOSITION

A wide range of 4-CP loading rates was tested when this compound was the sole carbon source (210–3150 mg L^{-1}) and also in the studies of cometabolism (525–3150 mg L^{-1}) in the presence of a wide range of phenol loads, as can be seen in Table 1. Ammonium sulphate and phosphoric acid were used as nitrogen and phosphorous sources, respectively. A COD:N:P:micronutrients ratio of 100:0.5:0.1:0.05 (w:w) was fixed. The mineral solution added as micronutrients supply consisted on $FeCl_3$, $CaCl_2$, KCl and $MgSO_4$.

7.2.3 ANALYTICAL METHODS

4-CP and phenol were analyzed by HPLC/UV (Prostar, Varian) using a C18 column as stationary phase (Microsorb MW-100-5) and a mixture of acetonitrile and H_2O (40:60, vol.) as mobile phase. The flow rate was maintained at 1.0 mL min^{-1} and a wavelength of 280 nm was used. Total organic carbon (TOC) was measured by an OI Analytical Model 1010 TOC apparatus. Total and volatile suspended solids (TSS and VSS) were determined according to the APHA Standard Methods. [31] SEM micrographs were obtained with a Hitachi S-3000N apparatus. Contribution of abiotic processes such as volatilization and adsorption onto biomass was evaluated. Adsorption of 4-CP was measured on biomass samples after extraction with Soxhelt following the US-EPA 8041 method. Tests of volatilization were performed under identical operating conditions to those employed in the biodegradation experiments but in the absence of biomass. The results reported were the average values from duplicate runs. In all the cases, the standard errors were lower than 10%.

TABLE 1: Treatable 4-CP and phenol loading rates (mg g^{-1} VSS day^{-1}) in different SBR arrangements

Reactor Cometabolic tests	Biomass	4-CP	Phenol
SBR-control	AS	20–60	20–120
SBR-*P. putida*	*P. putida*	20–60	120
SBR-bioaugmented	Bioaugmented AS	120	20–120
4-CP as sole carbon source			
SBR-control	AS	8–55	—
SBR-bioaugmented	Bioaugmented AS	8–120	—

7.3 RESULTS AND DISCUSSION

7.3.1 EFFECTS OF ACCLIMATION AND BIOAUGMENTATION DURING THE START-UP

Acclimation and bioaugmentation during the start-up period of a SBR treating a 4-CP load of 2.4 mg g^{-1} VSS day^{-1} with the addition of phenol as cosubstrate (3.8 mg g^{-1} VSS day^{-1}) were evaluated. Runs were carried out using non-adapted and acclimated biomass. The time-evolution of phenol and 4-CP concentrations during the aerobic phase of the start-up period of SBRs inoculated with different biomass is shown in Fig. 1. Whereas non-adapted AS did not show any biodegrading activity, both compounds were depleted after 8 h upon adaptation. Owing to the high tolerance of *P. putida* to the presence of toxic compounds, removal efficiencies higher than 35% were obtained for both phenol and 4-CP with non-adapted *P. putida* However, those removal efficiencies were greatly increased after acclimation, reaching values of 82 and 100% for 4-CP and phenol, respectively.

Regarding the effect of bioaugmentation, the use of non-adapted *P. putida* did not enhance the removal rates observed with adapted AS. This fact recommends the use of specialist degrading bacteria previously adapted to the toxic compounds for the bioaugmentation of bioreactors. The enhancement of the removal capacity of *P. putida* upon acclimation would improve the viability of the strain introduced. [32] The bioaugmented SBR showed the highest removal rates indicating the existence of synergic effects between the activated sludge and *P. putida*.

7.3.1 COMETABOLIC REMOVAL OF 4-CP
IN THE PRESENCE OF PHENOL

Figure 2 shows the time-evolution of phenol and 4-CP concentration during the aerobic reaction stage in a SBR inoculated with several adapted biomass. It is worth noting that phenol concentrations detected after filling in the control and bioaugmented SBR were lower than expected, which indicates that this compound was partially removed during the anoxic filling phase (Fig. 2(a)). With respect to 4-CP degradation, a long lag stage was observed in the SBR-*P. putida*, since 4-CP removal starts once phenol is almost depleted. The 4-CP removal rate increases sharply once phenol was nearly exhausted. This diauxic consumption of both compounds is commonly found in monoculture biological systems as well as negative interactions between chemically similar compounds in binary systems. [33] In this particular case, the initial step in the transformation of 4-CP is hydroxylation promoted by a NADPH-dependent monooxygenase giving rise to 4-chlorocatechol. The NADPH consumed in the monooxygenase reaction can be regenerated by the oxidation of phenol or phenol-induced biomass.34 Similarly, the oxidation of phenol is also initiated by a NADPH-dependent hydroxylase leading to catechol.

In contrast, simultaneous degradation of both substrates occurred in the control and bioaugmented SBR. It can also be seen that the bioaugmentation of AS with acclimated *P. putida* led to a significant increase in the removal rates. The bacterial pull present in these systems minimizes the competitive inhibition between 4-CP and phenol since different microorganisms can be involved in the degradation pathway of these compounds and the corresponding intermediates. Additionally, increasing phenol load up to 120 mg g^{-1} VSS day^{-1} allowed treatment of 4-CP loads up to 60 and 120 mg g^{-1} VSS day^{-1} in control and bioaugmented SBR, respectively. A possible explanation could be that phenol supplies the electrons required for the initial monooxygenase step of 4-CP biodegradation. [35] The high performance showed by the bioaugmented system can be caused by the exchange of genetic elements via cell-to-cell contact, which is thought to be significant for bacteria residing in aggregates representing high cell density environments. [36]

In all the experiments a residual fraction of TOC was measured in the resulting effluents, which was similar (20 mg L^{-1}) for control and bioaugmented SBR but almost triple in the case of SBR-*P. putida*. The remaining TOC has been related to the presence of soluble microbial products, whose concentration was significantly decreased by using AS as inoculum. Discounting that fraction, TOC removal was complete in all systems, reaching higher removal rates in bioaugmented SBR.

7.3.2 4-CP REMOVAL AS SOLE SUBSTRATE

The treatment of water with 4-CP as sole carbon and energy source by SBR inoculated with *P. putida* led to inactivation of the strain. Thus, the application of that single culture for the treatment of 4-CP in the absence of a cosubstrate is unlikely. Figure 3 shows the time-evolution of 4-CP concentration during the aerated stage in both non-bioaugmented (Fig. 3(a)) and bioaugmented (Fig. 3(b)) SBR. Although 4-CP disappearance rates were significantly lower than those achieved by the cometabolic systems, similar loads of 4-CP could be treated. The maximum 4-CP load (120 mg 4-CP g^{-1} VSS day^{-1}) treated efficiently by the bioaugmented SBR was considerably higher than for the SBR inoculated with AS. In this last system loads of 4-CP above 55 mg g^{-1} VSS day^{-1} caused a dramatic decrease of the removal efficiency, indicating that *P. putida* provides a greater resistance to toxicity. Although the decrease of indigenous AS concentration in the control SBR was negligible after acclimation to the toxic compound, it should not be ruled out that bioaugmentation of AS can exert a protective effect against the toxicity derived from the presence of 4-CP up to a certain concentration.

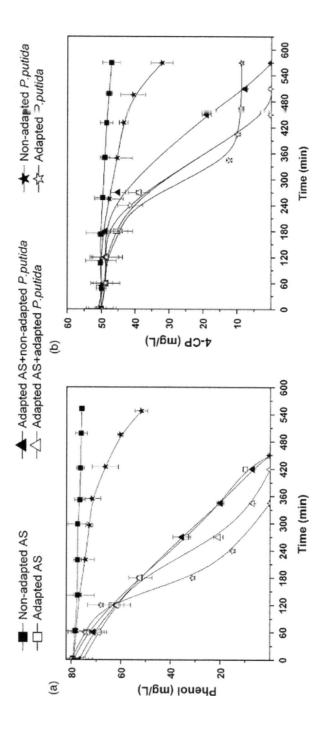

FIGURE 1: Time-evolution of phenol (a) and 4-CP (b) along the aerobic reaction stage of the start-up of SBR inoculated with different biomass.

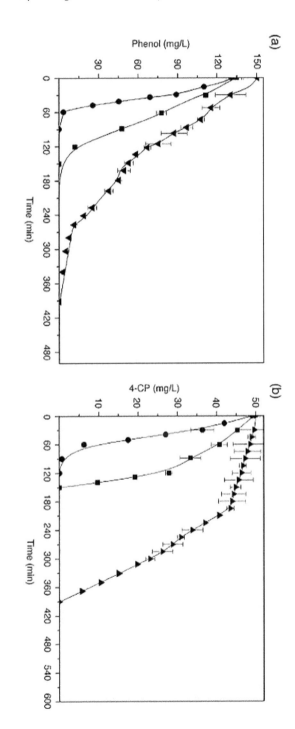

FIGURE 2: Time-evolution of phenol (a) and 4-CP (b) along the aerobic reaction stage in SBR inoculated with adapted AS (■), *P. putida* (triangle symbols) and AS + *P. putida* (●) treating 120 mg g⁻¹ VSS day⁻¹ of phenol and 40 mg g⁻¹ VSS day⁻¹ of 4-CP.

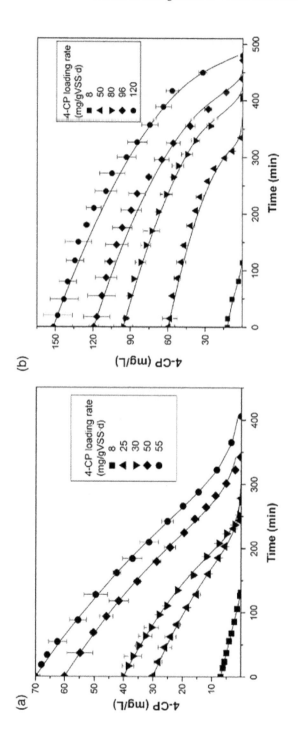

FIGURE 3: Time-evolution of 4-CP during the aerobic reaction stage of control (a) and bioaugmented (b) SBR treating 4-CP as sole substrate. Fittings to Haldane model (solid lines).

TABLE 2: Fitting values of the parameters* of Haldane model for 4-CP biodegradation as sole carbon source in control and bioaugmented SBR

	Control SBR					Bioaugmented SBR			
4-CP loading rates (mg g⁻¹ VSS day⁻¹)	Vmax (mg 4-CP g⁻¹ VSS min⁻¹)	KS (mg L⁻¹)	KI (mg L⁻¹)	r2	4-CP loading rates (mg g⁻¹ VSS day⁻¹)	Vmax (mg 4-CP g⁻¹ VSS min⁻¹)	KS (mg L⁻¹)	KI (mg L⁻¹)	r²
8	0.14 ± 0.01	1.28 ± 0.41	84.84 ± 41.22	0.999	8	0.60 ± 0.52	1.78 ± 0.74	1.86 ± 1.08	0.997
25	0.34 ± 0.06	5.57 ± 2.04	12.81 ± 3.59	0.999	50	1.26 ± 0.15	1.99 ± 1.37	2.31 ± 0.70	0.997
30	1.42 ± 0.41	29.48 ± 10.08	3.68 ± 1.16	0.999	80	4.35 ± 1.09	2.94 ± 1.09	2.81 ± 0.74	0.999
50	1.14 ± 0.24	39.45 ± 10.36	9.40 ± 2.27	0.999	100	2.37 ± 0.38	2.42 ± 1.03	7.96 ± 1.44	0.998
55	0.79 ± 0.11	27.97 ± 5.45	18.22 ± 3.13	0.999	120	5.14 ± 0.97	2.24 ± 0.72	5.02 ± 0.99	0.999

*Individual confidence limits at the 95% probability level.

The time-evolution of 4-CP concentration in both control and bioaugmented SBR showed inhibition profiles that were accurately described using the Haldane inhibition model (Fig. 3). The experimental data were fitted to the Haldane model. Fitting to the experimental results was accomplished using a non linear least squares minimization of the error using a simplex algorithm followed by a Powell minimization algorithm (Micromath® Scientist 3.0) at the 95% probability level to obtain values of the fitting parameters. The optimal-fitting values of the parameters (maximum specific 4-CP consumption rate, V_{max}, saturation, K_S, and inhibition, K_I, constants) for both systems are shown in Table 2. 4-CP loading rates above 30 mg 4-CP g^{-1} VSS day^{-1} led to a decrease of the maximum removal rate in the case of the control SBR, whereas in the bioaugmented SBR the value of this parameter continued increasing up to a load of 4-CP 2.5 times higher. The erratic trend of K_S and K_I values in the control SBR suggests the occurrence of changes in the microbial population. However, the saturation and inhibition constants increased when increasing the 4-CP concentration in the bioaugmented SBR. The increase of the K_I value could indicate a progressive acclimation of the microbial population to the presence of 4-CP. From this result it can be concluded that bioaugmentation is a more convenient intensification technique than acclimation for treating high toxic loads.

The bioindication analysis of the bioaugmented SBR after 3 months working showed the formation of 1.0–2.0 mm aggregates of free living bacteria (*P. putida*) caused by the presence of 4-CP in the medium (Fig. 4(a)). The toxic shock of 4-CP modified the cell morphology from bacillus to coccus, which led to spherical colonies in the bioaugmented SBR (Fig. 4(b)). The structural and morphological modifications seem to evolve towards a coccoid stress-resistant form, responsible for the residual viability of the microbial population. [37] In certain instances, not only the cell shape changes but also the volume of the cell decreases. [38]

The colonization process started with the surface adhesion of both *P. putida* and the bacteria of the AS. In a second step *P. putida* was covered with exopolysaccharides from the AS. The autoaggregation capacity and the inclusion of *P. putida* into the flocs may help to a successful bioaugmentation since it has been reported that non-flocculant strains did not survive in the AS system and required a higher inoculum size to accomplish substrate removal. [17]

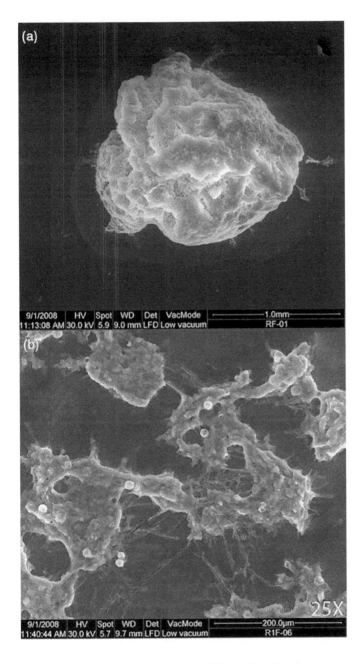

FIGURE 4: SEM micrographs of one aggregate of *P. putida* during the treatment of 4-CP in the SBR-*P. putida* (a) and bioaugmented activated sludge with *P. putida* (b).

7.4 CONCLUSIONS

The removal of 4-CP as representative of hardly biodegradable compounds by SBR was enhanced using cometabolism or bioaugmentation as intensification techniques. Acclimation of biomass improves significantly the extent and rate of 4-CP biodegradation during the start-up of the SBR. Furthermore, this period can be accelerated by bioaugmentation with specialist degrading bacteria. Bioaugmentation with *P. putida* enhanced the ability of this system to degrade 4-CP. Specialist degrading bacteria also must be adapted to the pollutant before being introduced in the SBR.

The addition of phenol as cosubstrate greatly enhanced the 4-CP removal rates at similar 4-CP loads. On the other hand, bioaugmentation with *P. putida* enhanced the ability of this system to degrade higher 4-CP loads. The biodegradation kinetics was satisfactorily described by the Haldane model. The morphology of *P. putida* changed from bacillus to coccus when 4-CP was added, giving rise to the formation of spherical colonies. These aggregates were integrated in the AS flocs, which favoured the retention of *P. putida* into the system. The high survival of this strain and a possible genetic transference with the microorganisms contained in the AS make the SBR a more suitable system for the biodegradation of recalcitrant pollutants.

REFERENCES

1. Annachhatre AP and Gheewala SH, Biodegradation of chlorinated phenolic compounds. Biotechnol Adv 14: 35–56 (1996).
2. Polo A, Tobajas M, Sanchis S, Mohedano AF and Rodríguez JJ, Comparison of experimental methods for determination of toxicity and biodegradability of xenobiotic compounds. Biodegradation 22: 751–761 (2011).
3. Puyol D, Mohedano AF, Sanz JL and Rodriguez JJ, Anaerobic biodegradation of 2,4,6-trichlorophenol by methanogenic granular sludge: role of co-substrates and methanogenic inhibition. Water Sci Technol 59: 1449–1456 (2009).
4. Konya I, Eker S and Kargi F, Mathematical modelling of 4-chlorophenol inhibition on COD and 4-chlorophenol removals in an activated sludge unit. J Hazard Mater 143: 233–239 (2007).
5. Kargi F and Konya I, COD, para-chlorophenol and toxicity removal from para-chlorophenol containing synthetic wastewater in an activated sludge unit. J Hazard Mater 132: 226–231 (2006).

6. Galíndez-Mayer J, Ramón-Gallegos J, Ruiz-Ordaz N, Juárez-Ramírez C, Salmerón-Alcocer A and Poggi-Varaldo HM, Phenol and 4-chlorophenol biodegradation by yeast Candida tropicalis in a fluidized bed reactor. Biochem Eng J 38: 147–157 (2008).

7. Kim J-H, Oh K-K, Lee S-T, Kim S-W and Hong S-I, Biodegradation of phenol and chlorophenols with defined mixed culture in shake-flasks and a packed bed reactor. Process Biochem 37: 1367–1373 (2002).

8. Zilouei H, Soares A, Murto M, Guieysse B and Mattiasson B, Influence of temperature on process efficiency and microbial community response during the biological removal of chlorophenols in a packed-bed bioreactor. Appl Microbiol Biotechnol 72: 591–599 (2006).

9. Monsalvo VM, Mohedano AF, Casas JA and Rodríguez JJ, Cometabolic biodegradation of 4-chlorophenol by sequencing batch reactors at different temperatures. Bioresource Technol 100: 4572–4578 (2009).

10. Pat AM, Vargas A and Buitron G, Practical automatic control of a sequencing batch reactor for toxic wastewater treatment. Water Sci Technol 63: 782–788 (2011).

11. Carucci A, Milia S, De Gioannis G and Piredda M, Acetate-fed aerobic granular sludge for the degradation of 4-chlorophenol. J Hazard Mater 166: 483–490 (2009).

12. Eker S and Kargi F, COD, para-chlorophenol and toxicity removal from synthetic wastewater using rotating tubes biofilm reactor (RTBR). Bioresource Technol 101: 9020–9024 (2010).

13. Eker S and Kargi F, Biological treatment of para-chlorophenol containing synthetic wastewater using rotating brush biofilm reactor. J Hazard Mater 135: 365–371 (2006).

14. Moreno-Andrade I, Buitron G and Vargas A, Effect of starvation and shock loads on the biodegradation of 4-chlorophenol in a discontinuous moving bed biofilm reactor. Appl Biochem Biotechnol 158: 222–230 (2009).

15. Vargas A, Moreno-Andrade I and Buitron G, Controlled backwashing in a membrane sequencing batch reactor used for toxic wastewater treatment. J Membrane Sci 320: 185–190 (2008).

16. Vargas A, Sandoval JL and Buitron G, Controlled operation of a membrane SBR for inhibitory wastewater treatment. Water Sci Technol 60: 655–661 (2009).

17. McLaughlin H, Farrell A and Quilty B, Bioaugmentation of activated sludge with two Pseudomonas putida strains for the degradation of 4-chlorophenol. J Environ Sci Health Part A Toxic/Hazard Subst Environ Eng 41: 763–777 (2006).

18. Chen Y, Cheng JJ and Creamer KS, Inhibition of anaerobic digestion process: a review. Bioresource Technol 99: 4044–4064 (2008).

19. Tobajas M, Monsalvo VM, Mohedano AF and Rodriguez JJ, Enhancement of cometabolic biodegradation of 4-chlorophenol induced with phenol and glucose as carbon sources by Comamonas testosteroni. J Environ Manag 95: S116–S121 (2012).

20. Buitron G and Moreno-Andrade I, Biodegradation kinetics of a mixture of phenols in a sequencing batch moving bed biofilm reactor under starvation and shock loads. J Chem Technol Biotechnol 86: 669–674 (2011).

21. Martínez-Hernández S, Olguín E, Gómez J and Cuervo-López F, Acetate enhances the specific consumption rate of toluene under denitrifying conditions. Archives Environ Contam Toxicol 57: 679–687 (2009).

22. Basu SK and Oleszkiewicz JA, Factors affecting aerobic biodegradation of 2-chlorophenol in sequencing batch reactors. Environ Technol 16: 1135–1143 (1995).

23. Van LH, Top EM and Verstraete W, Bioaugmentation in activated sludge: current features and future perspectives. Appl Microbiol Biotechnol 50: 16–23 (1998).

24. Loh K-C and Cao B, Paradigm in biodegradation using Pseudomonas putida- a review of proteomics studies. Enzyme Microbiol Technol 42: 1 12 (2008).

25. McClure NC, Fry JC and Weightman AJ, Survival and catabolic activity of natural and genetically engineered bacteria in a laboratory-scale activated-sludge unit. Appl Environ Microbiol 57: 366–373 (1991).

26. Boon N, Goris J, De VP, Verstraete W and Top EM, Bioaugmentation of activated sludge by an indigenous 3-chloroaniline-degrading Comamonas testosteroni strain, I2gfp. Appl Environ Microbiol 66: 2906–2913 (2000).

27. Wilderer PA, Rubio MA and Davids L, Impact of the addition of pure cultures on the performance of mixed culture reactors. Water Res 25: 1307–1313 (1991).

28. Mohan SV, Mohanakrishna G, Veer Raghavulu S and Sarma PN, Enhancing biohydrogen production from chemical wastewater treatment in anaerobic sequencing batch biofilm reactor (AnSBBR) by bioaugmenting with selectively enriched kanamycin resistant anaerobic mixed consortia. Int J Hydrogen Energy 32: 3284–3292 (2007).

29. Zhao L, Guo J, Yang J, Wang L and Ma F, Bioaugmentation as a tool to accelerate the start-up of anoxic-oxic process in a full-scale municipal wastewater treatment plant at low temperature. Int J Environ Pollut 37: 205–215 (2009).

30. Farrell A and Quilty B, Degradation of mono-chlorophenols by a mixed microbial community via a meta-cleavage pathway. Biodegradation 10: 353–362 (1999).

31. APHA, Standard Methods for the Examination of Water and Wastewater. American Public Health Association, Washington DC (1992).

32. Limbergen HV, Top EM and Verstraete W, Bioaugmentation in activated sludge: current features and future perspectives. Appl Microbiol Biotechnol 50: 16–23 (1998).

33. Reardon KF, Mosteller DC, Rogers JB, DuTeau NM and Kim K-H, Biodegradation kinetics of aromatic hydrocarbon mixtures by pure and mixed bacterial cultures. Environ Health Perspect 110:(2002).

34. Saéz PB and Rittmann BE, Biodegradation kinetics of a mixture containing a primary substrate (phenol) and an inhibitory co-metabolite (4-chlorophenol). Biodegradation 4: 3–21 (1993).

35. Chiavola A, McSwain BS, Irvine RL, Boni MR and Baciocchi R, Biodegradation of 3-chlorophenol in a sequencing batch reactor. J Environ Sci Health Part A 38: 2113–2123 (2003).

36. Wuertz S, Okabe S and Hausner M, Microbial communities and their interactions in biofilm systems: an overview. Water Sci Technol 49: 327–336 (2004).

37. Cefalì E, Patanè S, Arena A, Saitta G, Guglielmino S, Cappello S, et al, Morphologic variations in bacteria under stress conditions: near-field optical studies. Scanning 24: 274–283 (2002).

38. Fakhruddin A and Quilty B, Measurement of the growth of a floc forming bacterium Pseudomonas putida CP1. Biodegradation 18: 189–197 (2007).

CHAPTER 8

Treatment of Wastewaters from a Personal Care Products Factory by Advanced Biological Systems

VICTOR M. MONSALVO, JESUS LOPEZ, MIGUEL M. SOMER, ANGEL F. MOHEDANO, AND JUAN J. RODRIGUEZ

8.1 SHORT DESCRIPTION OF THE PROJECT

Cosmetic wastewaters are characterized by relatively high values of chemical oxygen demand (COD), suspended solids, fats, oils and detergents [1,2]. These effluents have been commonly treated by means of coagulation/flocculation [3,4]. Nevertheless, the more stringent regulations concerning industrial wastewaters makes necessary to implement new technologies. For this reason, the application of activated carbon adsorption [5], ultrafiltration [6] and advanced oxidation processes [7], including catalytic wet peroxide oxidation [8,9] has been reported in the last years.

In this work, we describe the treatment of wastewater from a cosmetic factory located in Spain. Initially, the treatment sequence consists of homogenization, filtration, coagulation, neutralization, flocculation, flota-

An original article, printed with permission from the authors.

tion and biological oxidation in a sequencing batch reactor (SBR). The main difficulties for treating cosmetic wastewater by biological processes derive from the presence of detergents, surfactants, hormones, cosmetics and pharmaceutical compounds [10]. Owing to capacity revamping and improvement of the piping, cleaning systems and process units in the last years, the organic loads have increased dramatically with a negative effect on the quality of the effluent. Given the high water needs and the policy adopted in the factory on water recycling as well as the existing limitations in land, a MBR was evaluated as a cost-efficient system.

8.2 DESIGN PARAMETERS

KOCH Membrane Systems Inc. used the following criteria for designing the ultrafiltration system of the revamped plant:

- Flexibility: The aqueous stream flow will vary within 70–360 m^3/d. The system uses a modular design, which enables part of the plant to be revamped.
- Quality control: The effluent from the plant has to be of proven quality. On-line monitoring and long-term logging of the water quality will ensure a reliable plant operation and supply.
- Ease of operation: Unlike conventional wastewater treatment plants, an ultrafiltration system cannot be operated manually. All flows and times have to be tightly controlled. Therefore full automatic operation was implemented.
- Minimisation of the environmental impact: Production of wastewater, energy consumption and chemicals needs have to be minimised.

Taking into account these premises, KOCH Membrane Systems Inc. designed the following MBR framework:

- The plant is fitted with two membrane modules in a single membrane tank separated from the biological reactor, providing a total filtration area of 1,000 m^2. Mixed liquor is recycled from the aeration reactor to the membranes tank at a feed/recycling ratio of 1:4 approximately.
- The membrane assembly is immersed in a 25 m^3 membranes tank. A floatable switch is also mounted in this tank to control the water level. The aeration system of the membranes tank was divided into two parts to provide two separate zones which were aerated alternatively in 60 s cycles.

Owing to the insufficient nutrients concentration in the raw wastewater an external nutrient source was used in order to reach a COD:N:P ratio of 100:5:1. For this reason, nutrients removal is not necessary and the biotreatment comprises just one aerobic reactor.

8.3 OPERATIONAL CRITERIA

The system is operated at an average permeate flux of 12 L/m²·h (LMH) working at a transmembrane pressure (TMP) of 272 ± 97 mbar. A filtration cycle of 6 min followed by a backflush of 30 s is used. A 30 s ventilation period is set every 20 cycles with a weekly cleaning cycle, using 1000 mg/L sodium hypochlorite and 2000 mg/L citric acid carried out by a clean-in-place (CIP) system, which takes 90 min. The intermittent air scouring of the PURON® module is conducted in repeated cycles of about 60 s combined with the filtration interval. To assist the flushing of the module, there is an additional backflush supplied after each filtration interval. The aeration frequency was 50% of the filtration time.

Coarse bubble aeration is applied at a rate of 0.2 Nm³/h·m³ of membrane area, operating at a specific air demand of 23 Nm³ air/m³ permeate. Hollow fibers are degasified for 120 s every 6 h. The permeability was 35.6 ± 8.9 LMH/bar over the range of operation TMP. Regular flushing, around once every 5 min, together with the CIP procedures outlined, enable the required steady-state flow rates to be maintained.

8.4 FLOW DIAGRAM OF THE PLANT

The full treatment scheme comprises a pre-treatment with coagulation and floating flocculation unit using polychlorinated aluminium as coagulant and a cationic flocculant (Fig. 1). This is followed by biological treatment in a stainless steel tank where two blowers provide 700 Nm³/h of air which maintains the DO within 1–5 mg/L. The MLSS in the bioreactor is kept between 10 and 14 g/L with a sludge production rate of 0.25 kg solids/kg COD.

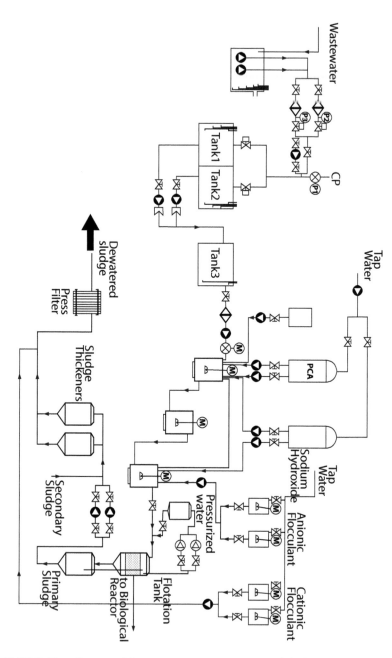

FIGURE 1: Flow diagram of the wastewater pre-treatment and sludge treatment system.

8.4.1 WASTEWATER STORAGE

Raw wastewater is homogenized in underground sealed storage tanks. Under regular continuous flow conditions the quarry pools level will fluctuate and drawing of each tank is automatically controlled.

8.4.2 PRE-TREATMENT

The coagulant and flocculant dosage is adjusted according to the jar-tests. Under automatic control the flocculant feed is regulated to guarantee quick and optimum flocculation.

8.4.3 CONTROL SYSTEM

The system offers the common flexibility and performance of a distributed control system (DCS) and it is designed for a minimum operator intervention and maintain the necessary interlocks for safe operation. To facilitate the control of the plant a PLC-based system has been selected with a PC-based topology incorporating a SCADA software package for easy operation.

8.4.4 CLEAN-IN-PLACE (CIP)

Chemical cleaning is achieved by pumping cleaning solutions from the CIP tank to the cell containing the membrane modules. The membranes are contacted with the chemicals and periodically aerated. At the completion of the cycle the cleaning solution is drawn from the system. The membranes are then rinsed and returned into service.

8.4.5 SCREENS

The incoming pre-treated wastewater prior to the membrane tank was passed through fine screening (3 mm).

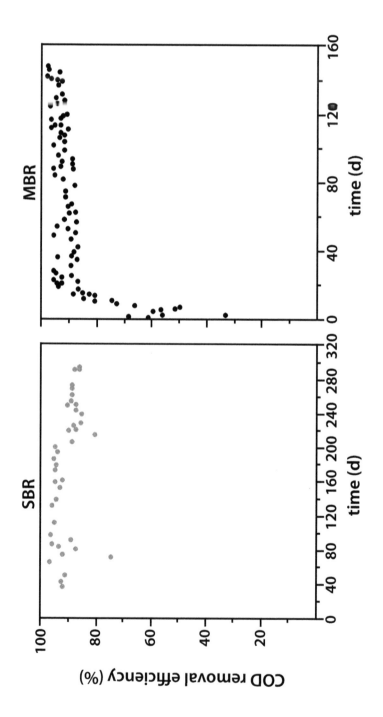

FIGURE 2: Time-evolution of COD removal efficiency by SBR and MBR.

8.4.6 BIOLOGICAL REACTOR AND MEMBRANES TANK

The activated sludge is recycled by pumping from the membranes tank into the bioreactor and then overflows back to the membranes tank. The filtration sequence comprises the following stages:

1. Standby. The system is kept in standby because either the level in the membranes tank is not sufficient or any critical element has failed. The module is aerated and the rest of elements are switched off.
2. Ventilation. During this stage, the air entrapped in the air trap is removed. This stage is performed automatically when the filtration sequence is started from standby. Permeate and ventilation valves are closed and opened, respectively, and the permeate pump generates a high back flow.
3. Filtration. Filtration valve is opened and permeate pump drives the permeate from the membranes tank to the permeate tank.
4. Backflush. When the filtration time is over, the permeate pump changes the flow direction working at a high rate for a given time.
5. De-aeration. Each certain number of cycles the permeate pump operates at a high flow in the filtration direction to remove the remaining air accumulated on top of the fibres.

8.4.7 SLUDGE TREATMENT

The dewatering unit consists of a sludge thickener combined with a press filter where primary and secondary sludges are treated together. The overall sludge production is ~ 3000 kg/d. A polymer dose of 0.57 kg/t DS is added, giving rise to an end product with a DS concentration of 30 wt%, which is disposed in a well-controlled landfill.

8.5 INFLUENT / EFFLUENT CHARACTERISTICS

Raw wastewaters contain a particulate and soluble COD, which is partially removed by the physico-chemical pre-treatment. BOD5, COD and solids removal efficiencies achieved more than 98% once the MBR system reached the steady state. A non-negligible FOG concentration is present in the raw wastewater which was also successfully removed (Table 1).

The quality of the effluent from the previous SBR system, which was discharged into the municipal sewer system, satisfied the limits regulated for industrial water discharges. However, due to the occurrence of filamentous bulking phenomenon, the settleability of the sludge suffered dramatic fluctuations reaching SVI values up to 400 mL/g SS. This caused that in some cases the solids concentration in the effluent were above the limits established by the water reuse legislation, thus hindering its reuse.

TABLE 1: Characteristics of the raw wastewater and the resulting effluent from the MBR plant.

Parameter	Influent	Effluent
BOD5 (mg O$_2$/L)	881	13
COD (mg O$_2$/L)	5,200	98
pH	7.1	8.3
Suspended solids (mg/L)	2,000	15
Conductivity (mS/cm)	1,601	2,600
Fats, oils and grease (mg/L)	234	< 10

The increasing organic load of the cosmetic wastewaters addressed to a slight decrease of the COD removal efficiency in the SBR (Fig. 2), reaching 86% when OLR was increased up to 0.6 kg COD/kg SS·d. It is somehow expected that a further increase of the OLR would cause a more severe drop on the SBR performance. Due to the acclimation of the biomass to the new operating conditions, a clear decrease of the COD removal efficiency was found during the start-up of the MBR. After a two weeks acclimation period, a noticeable increase of the COD removal efficiency occurred, reaching stable values close to 99%.

The presence of suspended solids in the effluent was substantially decreased, reaching concentrations in the permeate lower than 15 mg/L. Additionally, the values for turbidity, *Escherichia Coli* and *Legionella* satisfied the limits for industrial water reclamation, thereby providing the option of its reuse for general purposes such as irrigation and hosing down.

8.6 TECHNICAL CHALLENGES EXPERIENCED UPON IMPLEMENTATION

The plant has generally operated frankly well. The hydraulics of the system has never failed in this plant, and the membranes removed practically all suspended solids. In spite of the fact that disinfectant chemicals are discharged into the plant, the start-up of the biological reactor took short time since biomass from the former SBR system was kept in the new biological reactor.

The old air diffusion system was designed originally for the previous configuration to satisfy the oxygen consumption of the SBR which was run with a MLVSS concentration of 4,500 mg/L at the maximum. Owing to the increase of the MLVSS concentration in the MBR (8–10 g/L), prior to upgrading, the biological reactor was equipped with a more powerful aeration system with two blowers and membrane diffusers able to generate an air flow of 1,000 Nm^3/h.

There are some periods of wastewater scarcity due to a lower production in the factory. During these periods the plant is run by using the so-called "holidays mode". In order to maintain the plant working during such periods, the available volume of wastewater is stored and an inlet flow is set. Once holidays mode is activated, the wastewater feed is interrupted for 12 h; then wastewater is added every 24 h. The air blowers of the biological reactor are switched-on controlled by the PID depending on the dissolved oxygen concentration and the set point established. The filtration sequence is similar to the normal mode but in this case when the filtration is interrupted by the low level alarm, the blowers of the membranes tank do not run continuously but following programmed sequences.

8.7 MONITORING ISSUES

The visualisation system (man machine interface) runs on PCs with communication to the PLC. The PC runs a software package and represents the process configuration as MIMIC diagrams with individual displays of plant items or equipment and faceplates for control points. The combination of the visualisation and MMIC systems also offers trending, facilities, alarm handling, operator prompts and messages and data storage.

REFERENCES

1. J.A. Perdigón-Melón, J.B. Carbajo, A.L. Petre, R. Rosal, E. García-Calvo, Coagulation-Fenton coupled treatment for ecotoxicity reduction in highly polluted industrial wastewater, Journal of Hazardous Materials 181 (2010), 127-132.

2. D. Puyol, V.M. Monsalvo, A.F. Mohedano, J.L. Sanz, J.J. Rodriguez, Cosmetic wastewater treatment by upflow anaerobic sludge blanket reactor 185 (2010), 1059-1065.

3. H.G. Meiners, Purification of wastewater from cosmetic production, Parfuemerie und Kosmetik 75 (1994), 204-205.

4. G. Moggio, Pre-treatment implementation for industrial effluents from a cosmetic industry (Impianto di pretrattamento di effluenti industriali per una industria cosmética), Tecnol. Chim. 3 (2000), 60-64.

5. J.G. Catalán Lafuente, A. Bustos, J. Morá, J. Cabo, Study of the treatability of the waste water from a cosmetics factory (Estudio de la tratabilidad de aguas residuals de una industria cosmética), Doc. Invest. Hidrol. 18 (1974) 271–284.

6. I. Huisman, Optimising UF for wastewater treatment through membrane autopsy and failure analysis, Filtr. Sep. 41 (2004) 26–27.

7. P. Bautista, A.F. Mohedano, M.A. Gilarranz, J.A. Casas, J.J. Rodriguez, Application of Fenton oxidation to cosmetic wastewater treatment, J. Hazard. Mater. 143 (2007) 128–143.

8. P. Bautista, A.F. Mohedano, J.A. Casas, J.A. Zazo, J.J. Rodriguez, Oxidation of cosmetic wastewaters with H_2O_2 using Fe/-Al_2O_3 catalysts, Water Sci. Technol. 61 (6) (2010) 1631–1636.

9. P. Bautista, A.F. Mohedano, N. Menendez, J.A. Casas, J.J. Rodriguez, Catalytic wet peroxide oxidation of cosmetic wastewaters with Fe-bearing catalysts, Catal. Today 151 (2010) 148–152.

10. P.E. Stackelberg, E.T. Furlong, M.T. Meyer, S.D. Zaugg, A.K. Henderson and D.B. Reissman, Persistence of pharmaceutical compounds and other organic wastewater contaminants in a conventional drinking-water-treatment plant, Sci. Total Environ. 329 (2004), 99-113.

CHAPTER 9

Adsorption and Photocatalytic Decomposition of the β-Blocker Metoprolol in Aqueous Titanium Dioxide Suspensions: Kinetics, Intermediates, and Degradation Pathways

VIOLETTE ROMERO, PILAR MARCO, JAIME GIMÉNEZ, AND SANTIAGO ESPLUGAS

9.1 INTRODUCTION

The presence of pharmaceutical drugs and endocrine disruptors in surface, ground, and drinking waters is a growing environmental concern [1–9]. This pollution is caused by emission from production sites, direct disposal of surplus drugs in households, excretion after drug administration to humans and animals, wastewater from fish and other animal farms, and industry [3, 10, 11]. Some of these drugs, as β-blockers, have been detected in the order of $ng\,L^{-1}$ to $\mu g\,L^{-1}$ in the water [3–9, 12]. As an example, metoprolol tartrate salt (MET), which is usually prescribed as antihypertensive

Adsorption and Photocatalytic Decomposition of the -Blocker Metoprolol in Aqueous Titanium Dioxide Suspensions: Kinetics, Intermediates, and Degradation Pathways. © Romero V, Marco P, Giménez J, and Esplugas S. International Journal of Photoenergy **2013** (2013). http://dx.doi.org/10.1155/2013/138918. *Licensed under a Creative Commons Attribution 3.0 Unported License, http://creativecommons.org/licenses/by/3.0/.*

or antiarrhythmic, has been quantified up to $2\,\mu g\,L^{-1}$ in sewage treatment plant (STP) effluents and to $240\,ng\,L^{-1}$ in rivers [13]. Metoprolol and atenolol together account for more than 80% of total β-blockers consumption in Europe [6]. During the last years, metoprolol usage increased by a factor of 4, probably due to a change in human behavior [6]. Although full ecotoxicity data are not available [13, 14], it has been shown that they can adversely affect aquatic organisms, even at low concentration [2]. Due to its widespread occurrence and potential impact, MET must be removed from treated water before discharge or reuse.

Several treatments for the removal of these compounds have been reported in the literature, including membrane filtration [15], activated carbon adsorption [16], and reverse osmosis [17, 18]. However, the conventional water treatment processes are relatively inefficient in treating these compounds [4, 19]. These pharmaceuticals can undergo abiotic degradation (hydrolysis, photolysis) [13] and most of them are photoactive because their structural compositions consist of aromatic rings, heteroatoms, and other functional groups that can absorb solar radiation [20]. Thus, sunlight induced photochemical treatments should be considered as an alternative to traditional treatment. Several researches have demonstrated that MET shows slow direct phototransformation and/or hydrolysis [13, 21, 22]. In this context, advanced oxidation processes (AOPs) appear as a good alternative for its degradation due to their versatility and ability to increase biodegradability [23, 24]. Among the different advanced oxidation processes, heterogenous photocatalysis has been a potential alternative for the degradation of hazardous pollutants. Oxidation of organic compounds by means of TiO_2 was achieved by hydroxyl radical generation through the e^-/h^+ pair generated when the semiconductor is exposed to UV radiation [11, 14].

The main objective of this investigation is to undertake a study on the heterogeneous photocatalytic degradation and mineralization of MET in aqueous suspensions with TiO_2. In addition, the contribution of the degradation of MET by direct photolysis and the adsorption of the metoprolol onto TiO_2 were studied. In this way, the effect of different initial pH values on the photodegradation rate and the adsorption isotherms of metoprolol in TiO_2 suspensions were determined. The contribution of direct photolysis in photocatalysis was also examined in detail by using different wavelengths and glass type photoreactors. Additionally, an attempt has been completed to estimate the kinetic parameters and to identify the main intermediates formed during the photocatalytic degradation of MET.

9.2 MATERIALS AND METHODS

9.2.1 CHEMICALS AND REAGENTS

Metoprolol tartrate (MET) salt was purchased from Sigma Aldrich Chemical Co. (Spain) and used as received (1-[4-(2 methoxyethyl)phenoxy]-3-(propan-2-ylamino)propan-2-ol tartrate (2:1), CAS no 56392-17-7, $(C_{15}H_{25}NO_3)_2 \cdot C_4H_6O_6$, MW 684.81). Solutions of $50\,mg\,L^{-1}$ of MET were prepared using deionized water to assure accurate measurements of concentrations, to follow the TOC, to secure identification of intermediates, and to make predictions about possible mechanisms of photocatalysis. For pH adjustment, $0.1\,mol\,L^{-1}$ sulphuric acid or $0.1\,mol\,L^{-1}$ sodium hydroxide was used. All chemicals were HPLC grade, and they were used without further purification. Titanium dioxide (TiO_2) Degussa P-25 (commercial catalyst ~70% anatase, ~30% rutile, surface area $50 \pm 5.0\,m^2\,g^{-1}$, and $300\,\mu m$ particle size [25]) was used as received. This TiO_2 is a photochemical stable material [6, 26, 27].

9.2.2 ANALYTICAL INSTRUMENTS

The target compounds concentrations were monitored by a high-performance liquid chromatograph (HPLC) from Waters using a *SEA18* $5\,\mu m$ 15×0.46 Teknokroma column and Waters 996 photodiode array detector using Empower Pro software 2002 Water Co. The mobile phase was composed by water and acetonitrile (20:80), injected with a flow-rate of $0.85\,mL\,min^{-1}$, and detected at maximum metoprolol (221.9 nm). Total organic carbon (TOC) was measured in a Shimadzu TOC-V CNS. pH was measured by a Crison GLP 22 instrument. UV-VIS spectra of MET (Figure 1) were obtained for $10\,mg\,L^{-1}$ aqueous solution on a PerkinElmer UV/vis Lambda 20 (200–400 nm range) spectrophotometer.

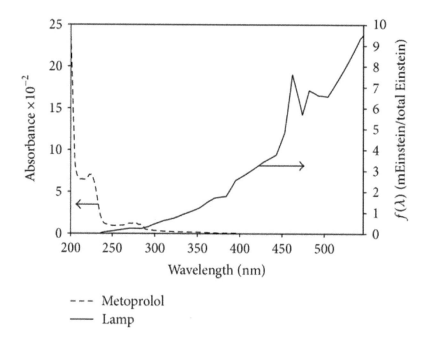

FIGURE 1: Absorbance spectrum of MET for aqueous concentration of $10\,\text{mg}\,\text{L}^{-1}$ (left axis) and lamp spectrum (right axis), where $f(\lambda)$ represents the spectral distribution of the lamp.

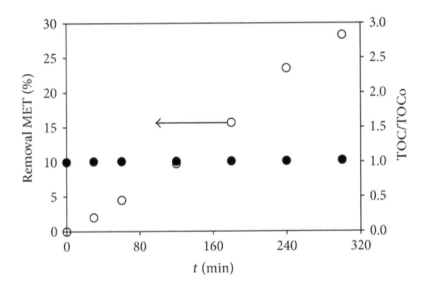

FIGURE 2: MET photodegradation removal (○) and TOC/TOCo (●) under simulated UV.

9.2.3 EXPERIMENTAL PROCEDURE

Photodegradation experiments were conducted in a Solarbox (CO. FO.MEGRA, Milan, Italy) and equipped with a Xenon lamp (Phillips XOP, 1000W) and a tubular-horizontal photoreactor (0.084L illuminated volume) located at the axis of a parabolic mirror in the bottom of the Solarbox. The photon flux inside the photoreactor was evaluated by o-nitrobenzaldehyde actinometry [28, 29], being 2.68 µEinstein s^{-1}. A stirred reservoir tank (1.0 L) was filled with the pharmaceutical-TiO$_2$ (suspended) aqueous solution. The solution was continuously pumped (peristaltic pump Ecoline VC-280 II, Ismatec) to the equipment and recirculated to the reservoir tank with a flow of 0.65 Lmin^{-1}. In order to keep the solution at 25°C, the jacket temperature of the stirred tank was controlled with an ultrathermostat bath (Haake K10). Samples were taken every 30 minutes during 300 minutes and quickly analyzed. Before HPLC analysis, samples were filtered through 0.20 μm PVDF membrane to separate TiO_2. All the experiments were duplicated and the results presented were the mean values.

According to the literature [13], metoprolol stability in aqueous solution was previously verified, by storing $50\,mg\,L^{-1}$ during 3 days in the dark at room temperature, and no degradation was observed.

MET adsorption of TiO_2 was also measured. Thus, MET solution (0 to $50\,mg\,L^{-1}$) was prepared with TiO2 in suspension ($0.4\,g\,L^{-1}$) and placed into 25 mL hermetic closed flasks, adjusting the pH with NaOH solution ($0.1\,mol\,L^{-1}$). The conical flasks were shaken at a constant speed of 100 rpm and at room temperature ($25 \pm 0.5°C$). Samples were taken every 24 h, assuming that adsorption equilibrium was reached.

For the identification of byproducts, the final sample mixture, at 300 minutes, was analyzed by electrospray ionization/mass spectrometry using a PerSeptive, TOF Mariner Jascoo AS-2050 plus IS mass spectrometer into the m/z range of 50–1000. The experiments were carried out in replicate.

9.3 RESULTS AND DISCUSSION

9.3.1 EFFECT OF UV RADIATION PHOTOLYSIS ON METOPROLOL DEGRADATION

When studying photocatalysis, it is very important to be able to separate the influence of photolysis, since it is expected to tackle the degradation of the substances mainly induced by the action of the catalyst. For this purpose, a series of experiments was done with UV illumination, and without catalyst to highlight the metoprolol ability to absorb the radiation reaching the system.

Figure 2 shows the results obtained after applying simulated sunlight. As observed, MET is not fast enough to be photodegraded in water by direct photolysis [30]; only 26% of MET in 300 minutes was degraded under simulated UV. Moreover, it shows that direct photolysis was not able to produce MET mineralization at the experimental conditions tested. This behavior can be explained because the MET absorption spectrum overlaps only slightly the spectrum of the incoming radiation (Figure 1).

The UV-VIS absorbance was used to calculate the molar absorption coefficient (ε) of the metoprolol at a wavelength of 221.9 nm (Figure 1), assuming that Beer-Lambert's law is followed:

$$A = -\log(T) = \varepsilon \times l \times C \tag{1}$$

where A is the absorbance (measured directly by the spectrophotometer), T is the transmittance, ε is the molar absorption coefficient, l is the distance that the light travels through the material, and C is the concentration of pollutant. The molar absorption coefficient (ε) was $281 \, \mathrm{L \, mol^{-1} \, cm^{-1}}$; this value is very similar to other reported values [12, 13]. This low value explains the MET stability in direct photolysis conditions. Nevertheless, different studies [12, 13, 31] show a high photoability of some β-blockers, for example, propranolol, nadolol, and alprenolol. The rapid photodegradation of these compounds was supported by a high molecular absorption coefficient ($\varepsilon > 800 \, \mathrm{L \, mol^{-1} \, cm^{-1}}$). This confirmed the hypothesis that photoinitiated reactions contribute to the degradation of naphthalene backbone (i.e., propranolol) [32], whereas the metoprolol, having a benzoic skeleton, is not sensitive to direct photolysis when dissolved in deionized water [33].

The efficiency of the photochemical transformation process depends on many factors such as the irradiation setups, the characteristics of the light source, the water matrix used, the initial concentration, and the pH of the solutions [31]. Tests were carried out using different photoreactors, a borosilicate Duran, and quartz glass reactor, cutting out wavelengths shorter than 290 and 320 nm, respectively. Other experiments have been done with and without glass filter for restricting transmissions of light below 280 nm.

In this study, the effect of borosilicate Duran and quartz glass material reactor has been investigated under UV radiation. It was observed that MET removal was 25% and 28% and the TOC reduction was 3.60% and 1.62%, for reactors made with borosilicate Duran and quartz, respectively.

Thus, although the mineralization was not significant, there is a small photodegradation of MET for the two tested reactors, after 300 minutes of reaction. Moreover, the effect of a filter glass cutting out wavelengths shorter than 280 nm has been investigated. As a result, only 19% of MET in 300 minutes has been removed with the glass filter; however, MET removal of 25% can be achieved without filter glass. TOC conversion was 6.37% and 3.60% with and without filter, correspondingly after 300 minutes, thus confirming that the mineralization is very low in both cases.

Summarizing, UV irradiation in the absence of TiO_2 achieved an MET degradation lower than 30% after 300 minutes of irradiation, confirming that the direct photolysis is not fast enough to be considered as an adequate technology.

9.3.2 THE ROLE OF THE ADSORPTION ON THE PHOTOCATALYTIC DEGRADATION

Since the adsorption can play an important role in the evolution of the photodegradation, adsorption experiments at constant temperature (25 ± 0.5°C) were carried out. The adsorption capacity of MET, q_e ($mg\,g^{-1}$), was calculated from the difference in MET concentration in the aqueous phase before and after adsorption at different initial MET concentrations (0, 6.2, 12.5, 25, 37.5, and 50 mg L^{-1}). The variation in adsorption of MET onto TiO_2 was studied at two pHs: 9 and free pH (pH ≈5.8). Figure 3 presents the obtained results and indicates that the amount adsorbed increases when pH does it.

The increase in the adsorption of metoprolol with increasing pH can be elucidated by considering the surface charge of the adsorbent material (pH_{pzc} ~ 6.5) [13, 34]. That is, titanium dioxide surface is positively charged in acid media pH (pH ≤ 7) whereas it is negatively charged under alkaline conditions (pH ≥ 7) [29, 35]. Also, metoprolol can be transformed to MET anion in the basic pH (pH ~ 10) since the pKa value of metoprolol is 9.7 [36]. Under free pH conditions, close to the point zero charge of TiO_2 (6.5) [30], MET is positively charged. A low adsorption was observed due to no electrostatic attraction between the surface charge and

MET. A highest adsorption between MET and TiO$_2$ would be observed at pH 9, because the negative charges of the surface of the catalyst attract the protonated MET form. In addition the photocatalytic degradation would be expected on the surface of the catalysis.

Two-parameter isotherm models (Langmuir, Freundlich, Temkin, and Dubinin-Radushkevich) and three-parameter isotherm models (Redlich-Peterson and Langmuir-Freundlich) were tested in the fitting of the adsorption data of MET onto titanium dioxide [37, 38].

K_a, K_p, K_T, K_{RP}, and K_L are the Langmuir, Freundlich, Temkin, Redlich-Peterson and Langmuir-Freundlich adsorption equilibrium constants (Lmol^{-1}), respectively; a_R and a_{IF} are also the Redlich-Peterson and Langmuir-Freundlcih constants (L mol^{-1}), respectively; C_e and q_e are the equilibrium concentration (mol L^{-1}) and the adsorption capacity (mol g^{-1}), respectively; Q_D is the Dubinin-Radushkevich saturation capacity (mol g^{-1}). The parameter q_m represents the maximum monolayer adsorption capacity (mol g^{-1}) and 1/n the adsorption intensity, which provides an indication of favorability and capacity of the adsorbent/adsorbate system. The parameter b is related to the adsorption heat; B_D gives the mean adsorption free energy E_D (kJ mol^{-1}). The parameters β and n_{LF} are the Redlich-Peterson and Langmuir-Freundlich exponents which lie between 0 and 1 [39]. And R^2 is the corresponding sum of squares error obtained in the fitting experimental data of each model.

From Table 1, it was observed that the best fitting were obtained for Langmuir isotherm ($R^2 = 0.987$ and 0.998 for free pH and pH 9, respectively). Thus, these models represent the equilibrium adsorption of MET on TiO$_2$ particles in the range of concentration studied. Accordingly, the adsorption mechanism may be interpreted as a monolayer coverage of the catalyst surface.

For free pH MET adsorption (q_m) was lower than for pH 9, 0.0014 mol g^{-1}, and 0.0250 mol g^{-1}, respectively. In these cases, the adsorption does not play an important role in the photocatalytic process. The MET percentage removal in dark conditions was 0.1% and 11% for free pH and pH 9. These low adsorption values and MET percentage removals suggest that the most possible way of degradation could be reached by migration of •OH radicals to the bulk of the suspension.

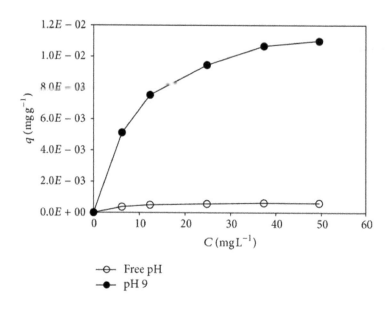

FIGURE 3: Effect of pH on adsorption of MET over TiO2 at 25°C, pH 9 (●), and free pH (○).

9.3.3 DEGRADATION OF MET BY PHOTOCATALYTIC PROCESS

The photocatalytic degradation of MET solution (50 mg L^{-1}) was carried out during 300 minutes, in the presence of 0.4 g L^{-1} of TiO$_2$ under UV-VIS light at room temperature. It is know that, in heterogeneous photocatalysis, the rate of degradation is not always proportional to the catalyst load [40]. An optimal point exists where TiO$_2$ loaded shows a maximum degradation rate. Previous studies carried out in our research group reported that the optimum catalyst concentration was 0.4 g L^{-1} [41]. Over this value, scattering can appear, and therefore increase in degradation rate does not occur.

Firstly, the solution mixture was stirred for 24 hours without irradiation in order to get the equilibrium of MET adsorption.

TABLE 1: Isotherm parameters for MET adsorption onto TiO_2 obtained by linear method at 25°C.

Two-parameter model	Parameters	pH	
		Free	9
Langmuir	q_m (mol g^{-1})	0.0014	0.0250
	K_L (L mol^{-1})	0.0930	0.0817
$q_e = \dfrac{q_m K_a C_e}{1 + K_a C_e}$	R^2	0.987	0.998
Freundlich	$1/n$	0.244	0.670
	K_F (L mol^{-1})	0.00042	0.00129
$q_e = K_F C_e^{1/n}$	R^2	0.105	0.147
Temkin	RT/b	0.00024	0.00508
	K_T (L mol^{-1})	2.718	0.962
$q_e = \dfrac{RT}{b} \ln(K_T + C_e)$	R^2	0.604	0.982
Dubinin-Radushkevich	q_D (mol g^{-1})	0.0013	0.0233
	$B_D \times 10^{-3}$ (mol^2 kJ^{-2})	1.533	1.532
$q_e = q_D \exp\left(-B_D \left[RT \ln\left(1 + \dfrac{1}{C_e}\right)\right]\right)$	R^2	0.835	0.778
Three-parameter model			
Redlich-Peterson	K_{RP} (L mol^{-1})	0.00011	0.10
	a_R (L mol^{-1})	0.051	18.040
$q_e = \dfrac{K_{RP} C_e}{1 + a_R C_e^{\beta}}$	β	0.999	0.679
	R^2	0.826	0.789
Langmuir-Freundlich	K_{LF} (L mol^{-1})	0.000205	0.00809
	a_{LF} (L mol^{-1})	0.134	0.310
$q_e = \dfrac{K_{LF} C_e^{n_{LF}}}{1 + (a_{LF} C_e)^{n_{LF}}}$	n_{LF}	0.779	0.92
	R^2	0.211	0.734

Figure 4 depicts the photocatalytic degradation of MET at free pH and pH 9. Maximum conversions are achieved at 240 and 300 minutes for pH 9 and free pH, respectively. An important remark is that the initial removal rate for free pH and pH 9 experiments is different, being higher at pH 9. The effect of pH on the conversion is a complex issue related to the ionization states of the catalyst surface and the substrate, as well as the rate of formation of radicals and other reactive species in the reaction mixture [42]. These effects can be assessed since the action of the holes is favored at acidic conditions, while hydroxyl radicals become the dominant species at neutral and alkaline conditions [42].

As known, photocatalysis occurs through the energy adsorption by the catalyst (light between 200 and 400 nm for TiO_2). Under excited condition, the valance band-electron is transferred to the conduction band forming the hole-electron pair (h^+/e^-) (2). The hydroxyl radicals are formed by cleavage of adsorbed molecules of water [43]:

$$TiO_2 + h\gamma \rightarrow TiO_2^* \left(\frac{h^+}{e^-}\right) \tag{2}$$

$$h^+ + H_2O \rightarrow \bullet OH + H^+ \tag{3}$$

If organic compounds are absorbed on the surface of the catalyst, the •OH nonselective attack promotes the cleavage of compounds bounds. The higher MET degradation and the low MET adsorption on catalyst at a pH 9 suggest that the •OH attack in the bulk of solution can be responsible for the MET degradation [40, 44].

The values of TOC during the photocatalytic degradation of MET, at two different pHs, are given in Figure 4. The TOC increases with time, indicating the increasing mineralization of the initial organic structures.

9.3.4 KINETICS OF MET DEGRADATION

The Langmuir-Hinshelwood (L-H) model is usually used to describe the kinetics of photocatalytic degradation of organic pollutants [12, 13, 30, 31, 41, 45], being the kinetic equation expressed as

$$r = -\frac{dC}{dt} = \frac{K_{ads} \cdot K_{H-L} \cdot C}{1 + K_{ads} \cdot C} \tag{4}$$

where r is the degradation rate, C is the reactant concentration, t is the time, k_{H-L} is the rate constant, and k_{ad} is the adsorption equilibrium constant.

This model assumes that adsorption is a rapid equilibrium process and that the rate-determining step of the reaction involves the species present in a monolayer at the solid–liquid interface. Furthermore if the adsorption of MET onto the surface of the photocatalysts is very low, $K_{ads} \bullet C$ can be neglected in the denominator simplifying the equation to a pseudo-first-order equation as given by [46]

$$r = -\frac{dC}{dt} = K_{ads} \cdot K_{H-L} \cdot C = k \cdot C \tag{5}$$

The integrated form of the above equation is represented by

$$\ln\left(\frac{C_0}{C}\right) = K_{app} \cdot t \tag{6}$$

where C_0 is the initial pollutant concentration and K_{app} is the apparent pseudo-first-order reaction rate constant. The half-life was calculated with the following expression:

$$t_{1/2} = \frac{\ln 2}{k} \tag{7}$$

The values of $t_{1/2}$ in Table 2 verify that the direct photolysis under simulated light was very low. The low photodegradation of MET was also supported by a low molar absorption coefficient ($281\, L\, mol^{-1}\, cm^{-1}$) measured at $221.9\, nm$ wavelength. However, an important increasing difference is observed in the MET degradation when TiO_2 is present. Also, when photocatalytic process is applied, results in TOC conversion (63%) are notoriously improved, for initial concentration of $50\, mg\, L^{-1}$ of MET and $0.4\, g\, L^{-1}$ of catalyst.

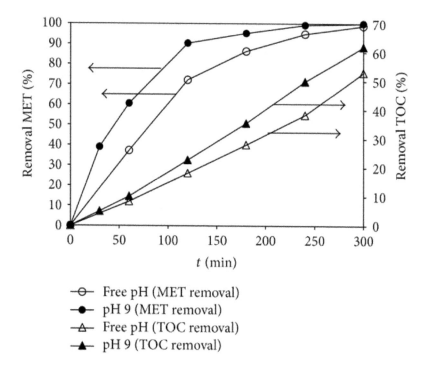

FIGURE 4: MET and TOC removal (%) versus time (min) at free pH and pH 9 in photocatalytic experiments.

TABLE 2: Kinetics of metoprolol UV-C photodegradation under different conditions.

	Glass material reactor	Glass filter λ ≥ 280 nm	pH	$t_{1/2}$ (h)
Photolysis	Borosilicate	with	5.8 ± 1	16.5 ± 0.5
		without	5.8 ± 1	11.6 ± 0.6
	Quartz	without	5.8 ± 1	10.5 ± 0.5
Photocatalysis	Borosilicate	without	5.8 ± 1	0.81 ± 0.4
		without	9.0 ± 1	0.58 ± 0.3

If both processes are compared, photocatalytic process is always much faster than the photolytic degradation of MET. Therefore, the interest of using photocatalysis in the treatment of this type of pollutant is obvious.

9.3.5 INTERMEDIATES DURING REACTION

The major by-products formed during 6 hours of photocatalytic treatment of MET were identified (Figure 5). The study was carried out using HPLC/MS in positive electrospray model. The degradation intermediates for MET are shown in Table 3.

The metoprolol has a molecular weight [M + H+] = 268. Three intermediates corresponding to the binding of •OH radicals in the aromatic ring were detected at m/z 300, 316, and 332, di-(DP (Detected Compound) 16), tri-(DP 17), and tetrahydroxy (DP 18) DPs, respectively. After breaking the C–C bond in the aliphatic part of the MET molecule, amino-diol (DP 8) was identified as one of the dominant intermediates with m/z = 134. Different fragments of the ethanolamine side were also identified (DP 1, DP 2, DP 3, DP 4, DP 5, DP 6, and DP 7), probably due to the loss of the hydroxyl group and the loss of isopropyl moiety.

PD 15 can be formed probably by reactions which involve attack on the ether side chain followed by elimination. On the other hand, the oxidation of alcohols to aldehydes can be explained by the formation of DP 14 with m/z = 238 [47]. The hydrogen abstraction and the water elimination of DP 14 probably generate a carbonyl, followed by an intermolecular electron transfer; it generates a double bond and the consequent formation of DP 12.

TABLE 3: Intermediates proposed for the photocatalytic degradation of MET.

Detected compound (DP)	Ret. time (min)	m/z (Da)	Molecular formula	Proposed structure
Metoprolol	3.24	268	$C_{15}H_{25}NO_3$	
1	2.59	74	$C_4H_{11}N$	
2	3.24	76	C_3H_9NO	
3	2.59	92	$C_3H_9NO_2$	
4	2.59	102	$C_5H_{11}NO$	
5	2.59	116	$C_6H_{13}NO$	
6	3.24	118	$C_6H_{15}NO$	
7	2.59	120	$C_5H_{125}NO_2$	
8	2.59	134	$C_6H_{15}NO_2$	
9	3.23	161	$C_{11}H_{12}O$	
10	3.23	193	$C_{12}H_{16}O_2$	
11	3.24	208	$C_{12}H_{17}NO_2$	

TABLE 3: *Cont.*

Detected compound (DP)	Ret. time (min)	m/z (Da)	Molecular formula	Proposed structure
12	3.03	220	$C_{13}H_{17}NO_2$	
13	3.03	226	$C_{12}H_{19}NO_3$	
14	4.71	238	$C_{13}H_{19}NO_3$	
15	3.23	240	$C_{13}H_{21}NO_3$	
16	3.98	300	$C_{15}H_{25}NO_5$	
17	3.98	316	$C_{13}H_{25}NO_6$	
18	3.23	332	$C_{13}H_{19}NO_7$	

Oxidative attack on the dimethylamine moiety results in a DP 13 with $m/z = 226$. Following this, the hydrogen abstraction and elimination of water of DP 13 generate a carbonyl which followed by intermolecular electron transference, generates a double bond and forming DP 11. The DP 11 can generate DP 10 corresponding to a loss of ammonia after the hydrogen abstraction. The intermediate 9 could be formed by the loss of methanol combined with the attack of •OH on the C atom next to the ether oxygen in the aliphatic part of DP 10. A simplified fragmentation pathway of metoprolol degradation is shown in Figure 6.

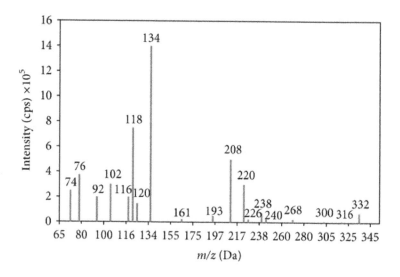

FIGURE 5: MS spectrum of major oxidation products of metoprolol.

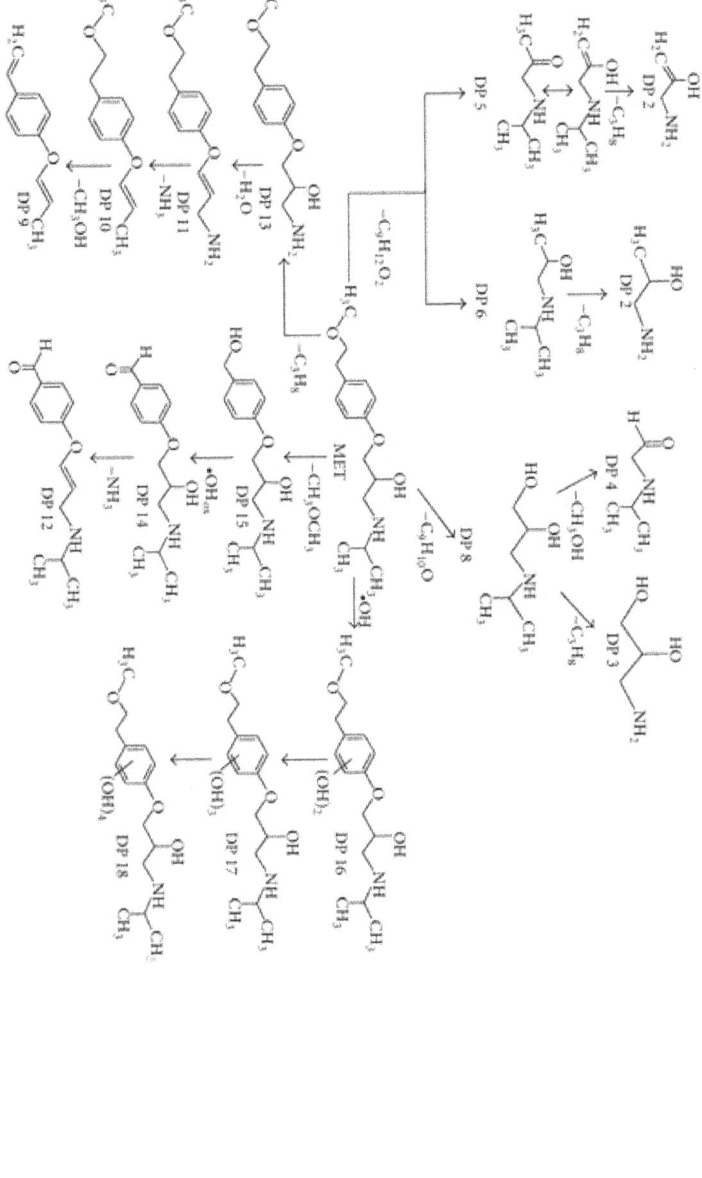

FIGURE 6: Proposed pathways for the degradation of MET.

9.4 CONCLUSIONS

Langmuir isotherm fits very well the experimental data, which indicates that the adsorption of the MET onto TiO_2 is by monolayer coverage of the catalyst surface. The results confirmed that the degradation of MET is not able to undergo by direct photolysis due to its lower absorption coefficient. In contrast, the addition of TiO_2 photocatalyst significantly increases its degradation rate and, after 240 min of irradiation, MET was totally eliminated for pH 9. The experimental data indicates that TiO_2 photocatalysis allows a fast and efficient removal of metoprolol, transforming substrate into by-products that are more difficult to be degraded by photocatalysis, as evidenced by the level of mineralization achieved (63%). Disappearance of MET by photocatalysis follows Langmuir-Hinshelwood model that can be simplified as a pseudo-first-order equation, as usually found in heterogenous photocatalysis at low concentration. Photocatalytic degradation rate of MET depends on pH, occurring the faster degradation at pH 9. At last, based on the identified degradation intermediates at 6-hour reaction time, a photocatalytic degradation pathway of metoprolol was proposed. The main pathways involved in the photocatalytic degradation process include hydroxilation of the aromatic ring, shortening of methoxyl contained in the lateral chain, and cleavage of or addition of •OH to the amine lateral chain.

REFERENCES

1. A. Pal, K. Y.-H. Gin, A. Y.-C. Lin, and M. Reinhard, "Impacts of emerging organic contaminants on freshwater resources: review of recent occurrences, sources, fate and effects," Science of the Total Environment, vol. 408, no. 24, pp. 6062–6069, 2010.
2. Y. Xu, T. V. Nguyen, M. Reinhard, and K. Y.-H. Gin, "Photodegradation kinetics of p-tert-octylphenol, 4-tert-octylphenoxy-acetic acid and ibuprofen under simulated solar conditions in surface water," Chemosphere, vol. 85, no. 5, pp. 790–796, 2011.
3. A. Jurado, E. Vàzquez-Suñé, J. Carrera, M. López de Alda, E. Pujades, and D. Barceló, "Emerging organic contaminants in groundwater in Spain: a review of sources, recent occurrence and fate in a European context," Science of the Total Environment, vol. 440, pp. 82–94, 2012.

4. M. Huerta-Fontela, M. T. Galceran, and F. Ventura, "Occurrence and removal of pharmaceuticals and hormones through drinking water treatment," Water Research, vol. 45, no. 3, pp. 1432–1442, 2011.

5. M. Pedrouzo, F. Borrull, E. Pocurull, and R. M. Marcé, "Presence of pharmaceuticals and hormones in waters from sewage treatment plants," Water, Air, and Soil Pollution, vol. 217, no. 1–4, pp. 267–281, 2011.

6. A. C. Alder, C. Schaffner, M. Majewsky, J. Klasmeier, and K. Fenner, "Fate of β-blocker human pharmaceuticals in surface water: comparison of measured and simulated concentrations in the Glatt Valley Watershed, Switzerland," Water Research, vol. 44, no. 3, pp. 936–948, 2010.

7. M. Maurer, B. I. Escher, P. Richle, C. Schaffner, and A. C. Alder, "Elimination of β-blockers in sewage treatment plants," Water Research, vol. 41, no. 7, pp. 1614–1622, 2007.

8. B. Abramović, S. Kler, D. Šojić, M. Laušević, T. Radović, and D. Vione, "Photocatalytic degradation of metoprolol tartrate in suspensions of two TiO2-based photocatalysts with different surface area. Identification of intermediates and proposal of degradation pathways," Journal of Hazardous Materials, vol. 198, pp. 123–132, 2011.

9. M. Maurer, B. I. Escher, P. Richle, C. Schaffner, and A. C. Alder, "Elimination of β-blockers in sewage treatment plants," Water Research, vol. 41, no. 7, pp. 1614–1622, 2007.

10. E. Isarain-Chávez, J. A. Garrido, R. M. Rodríguez et al., "Mineralization of metoprolol by electro-fenton and photoelectro-fenton processes," Journal of Physical Chemistry A, vol. 115, no. 7, pp. 1234–1242, 2011.

11. L. Prieto-Rodríguez, I. Oller, N. Klamerth, A. Agüera, E. M. Rodríguez, and S. Malato, "Application of solar AOPs and ozonation for elimination of micropollutants in municipal wastewater treatment plant effluents," Water Research, vol. 47, pp. 1521–1528, 2013.

12. I. Kim and H. Tanaka, "Photodegradation characteristics of PPCPs in water with UV treatment," Environment International, vol. 35, no. 5, pp. 793–802, 2009.

13. A. Piram, A. Salvador, C. Verne, B. Herbreteau, and R. Faure, "Photolysis of β-blockers in environmental waters," Chemosphere, vol. 73, no. 8, pp. 1265–1271, 2008.

14. H. Fang, Y. Gao, G. Li et al., "Advanced oxidation kinetics and mechanism of preservative propylparaben degradation in aqueous suspension of TiO2 and risk assessment of its degradation products," Environmental Science and Technology, vol. 47, pp. 2704–2712, 2013.

15. L. D. Nghiem, A. I. Schäfer, and M. Elimelech, "Pharmaceutical retention mechanisms by nanofiltration membranes," Environmental Science and Technology, vol. 39, no. 19, pp. 7698–7705, 2005.

16. T. Heberer, "Occurrence, fate, and removal of pharmaceutical residues in the aquatic environment: a review of recent research data," Toxicology Letters, vol. 131, no. 1-2, pp. 5–17, 2002.

17. C. Hartig, M. Ernst, and M. Jekel, "Membrane filtration of two sulphonamides in tertiary effluents and subsequent adsorption on activated carbon," Water Research, vol. 35, no. 16, pp. 3998–4003, 2001.

18. J. Radjenović, M. Petrović, F. Ventura, and D. Barceló, "Rejection of pharmaceuticals in nanofiltration and reverse osmosis membrane drinking water treatment," Water Research, vol. 42, no. 14, pp. 3601–3610, 2008.

19. P. Westerhoff, Y. Yoon, S. Snyder, and E. Wert, "Fate of endocrine-disruptor, pharmaceutical, and personal care product chemicals during simulated drinking water treatment processes," Environmental Science and Technology, vol. 39, no. 17, pp. 6649–6663, 2005.

20. J. Peuravuori and K. Pihlaja, "Phototransformations of selected pharmaceuticals under low-energy UVA-vis and powerful UVB-UVA irradiations in aqueous solutions-the role of natural dissolved organic chromophoric material," Analytical and Bioanalytical Chemistry, vol. 394, no. 6, pp. 1621–1636, 2009.

21. Q.-T. Liu, R. I. Cumming, and A. D. Sharpe, "Photo-induced environmental depletion processes of β-blockers in river waters," Photochemical and Photobiological Sciences, vol. 8, no. 6, pp. 768–777, 2009.

22. Q.-T. Liu, R. I. Cumming, and A. D. Sharpe, "Photo-induced environmental depletion processes of β-blockers in river waters," Photochemical and Photobiological Sciences, vol. 8, no. 6, pp. 768–777, 2009.

23. R. Molinari, F. Pirillo, V. Loddo, and L. Palmisano, "Heterogeneous photocatalytic degradation of pharmaceuticals in water by using polycrystalline TiO2 and a nanofiltration membrane reactor," Catalysis Today, vol. 118, no. 1-2, pp. 205–213, 2006.

24. W. Song, W. J. Cooper, S. P. Mezyk, J. Greaves, and B. M. Peake, "Free radical destruction of β-blockers in aqueous solution," Environmental Science and Technology, vol. 42, no. 4, pp. 1256–1261, 2008.

25. M. Janus, J. Choina, and A. W. Morawski, "Azo dyes decomposition on new nitrogen-modified anatase TiO2 with high adsorptivity," Journal of Hazardous Materials, vol. 166, no. 1, pp. 1–5, 2009.

26. T. E. Doll and F. H. Frimmel, "Fate of pharmaceuticals—photodegradation by simulated solar UV-light," Chemosphere, vol. 52, no. 10, pp. 1757–1769, 2003.

27. M. Šcepanovic, B. Abramovic, A. Golubovic et al., "Photocatalytic degradation of metoprolol in water suspension of TiO2 nanopowders prepared using sol-gel route," Journal of Sol-Gel Science and Technology, vol. 61, pp. 390–402, 2012.

28. K. L. Willett and R. A. Hites, "Chemical actinometry: using o-Nitrobenzaldehyde to measure light intensity in photochemical experiments," Journal of Chemical Education, vol. 77, no. 7, pp. 900–902, 2000. View at Scopus

29. N. De la Cruz, V. Romero, R. F. Dantas et al., "o-Nitrobenzaldehyde actinometry in the presence of suspended TiO2 for photocatalytic reactors," Catalysis Today, vol. 209, pp. 209–214, 2013.

30. F. J. Rivas, O. Gimeno, T. Borralho, and M. Carbajo, "UV-C radiation based methods for aqueous metoprolol elimination," Journal of Hazardous Materials, vol. 179, no. 1–3, pp. 357–362, 2010.

31. D. Fatta-Kassinos, M. I. Vasquez, and K. Kümmerer, "Transformation products of pharmaceuticals in surface waters and wastewater formed during photolysis and advanced oxidation processes - Degradation, elucidation of byproducts and assessment of their biological potency," Chemosphere, vol. 85, no. 5, pp. 693–709, 2011.

32. S. Sortino, S. Petralia, F. Boscà, and M. A. Miranda, "Irreversible photo-oxidation of propranolol triggered by self-photogenerated singlet molecular oxygen," Photochemical and Photobiological Sciences, vol. 1, no. 2, pp. 136–140, 2002.

33. Q.-T. Liu and H. E. Williams, "Kinetics and degradation products for direct photoly-
     sis of β-blockers in water," Environmental Science and Technology, vol. 41, no. 3,
     pp. 803–810, 2007.
34. P. Fernández-Ibáñez, F. J. De Las Nieves, and S. Malato, "Titanium dioxide/electro-
     lyte solution interface: electron transfer phenomena," Journal of Colloid and Inter-
     face Science, vol. 227, no. 2, pp. 510–516, 2000.
35. A. Mills and S. Le Hunte, "An overview of semiconductor photocatalysis," Journal
     of Photochemistry and Photobiology A, vol. 108, no. 1, pp. 1–35, 1997. View at
     Scopus
36. F. J. Benitez, J. L. Acero, F. J. Real, G. Roldan, and F. Casas, "Bromination of select-
     ed pharmaceuticals in water matrices," Chemosphere, vol. 85, no. 9, pp. 1430–1437,
     2011.
37. K. Y. Foo and B. H. Hameed, "Insights into the modeling of adsorption isotherm
     systems," Chemical Engineering Journal, vol. 156, no. 1, pp. 2–10, 2010.
38. K. V. Kumar and K. Porkodi, "Relation between some two- and three-parameter
     isotherm models for the sorption of methylene blue onto lemon peel," Journal of
     Hazardous Materials, vol. 138, no. 3, pp. 633–635, 2006.
39. J. S. Piccin, G. L. Dotto, and L. A. A. Pinto, "Adsorption isotherms and thermo-
     chemical data of FDandC RED N° 40 Binding by chitosan," Brazilian Journal of
     Chemical Engineering, vol. 28, no. 2, pp. 295–304, 2011.
40. F. Méndez-Arriaga, J. Gimenez, and S. Esplugas, "Photolysis and TiO2 photocata-
     lytic treatment of naproxen: degradation, mineralization, intermediates and toxic-
     ity," Journal of Advanced Oxidation Technologies, vol. 11, no. 3, pp. 435–444, 2008.
41. V. Romero, N. de La Cruz, R. F. Dantas, P. Marco, J. Giménez, and S. Esplugas,
     "Photocatalytic treatment of metoprolol and propranolol," Catalysis Today, vol. 161,
     no. 1, pp. 115–120, 2011.
42. L. A. Ioannou, E. Hapeshi, M. I. Vasquez, D. Mantzavinos, and D. Fatta-Kassinos,
     "Solar/TiO2 photocatalytic decomposition of β-blockers atenolol and propranolol in
     water and wastewater," Solar Energy, vol. 85, no. 9, pp. 1915–1926, 2011.
43. F. Méndez-Arriaga, S. Esplugas, and J. Giménez, "Photocatalytic degradation of
     non-steroidal anti-inflammatory drugs with TiO2 and simulated solar irradiation,"
     Water Research, vol. 42, no. 3, pp. 585–594, 2008.
44. H. Yang, T. An, G. Li et al., "Photocatalytic degradation kinetics and mecha-
     nism of environmental pharmaceuticals in aqueous suspension of TiO2: a case of
     β-blockers," Journal of Hazardous Materials, vol. 179, no. 1–3, pp. 834–839, 2010.
45. R. F. Dantas, O. Rossiter, A. K. R. Teixeira, A. S. M. Simões, and V. L. da Silva, "Di-
     rect UV photolysis of propranolol and metronidazole in aqueous solution," Chemi-
     cal Engineering Journal, vol. 158, no. 2, pp. 143–147, 2010.
46. C. Sahoo, A. K. Gupta, and I. M. S. Pillai, "Heterogeneous photocatalysis of real
     textile wastewater: evaluation of reaction kinetics and characterization," Journal of
     environmental science and health A, vol. 47, pp. 2109–2119, 2012.
47. M. L. Wilde, W. M. M. Mahmoud, K. Kümmerer, and A. F. Martins, "Oxidation-
     coagulation of β-blockers by K2FeVIO4 in hospital wastewater: assessment of
     degradation products and biodegradability," Science of the Total Environment, vol.
     452-453, pp. 137–147, 2013.

CHAPTER 10

Ozone Oxidation of Antidepressants in Wastewater: Treatment Evaluation and Characterization of New By-Products by LC-QToFMS

ANDRÉ LAJEUNESSE, MIREILLE BLAIS, BENOÎT BARBEAU, SÉBASTIEN SAUVÉ, AND CHRISTIAN GAGNON

10.1 BACKGROUND

Urban wastewaters are one of the major sources of pharmaceutically-active compounds (PhACs) into aquatic environments [1,2]. The elimination of many pharmaceuticals in sewage treatment plants (STPs) being often incomplete [3-5], effluents from STPs thus contribute to a significant load of pharmaceutical residues in the receiving waters [6]. Little is however known on the potential release of transformation by-products following advanced wastewater treatments.

Among the most prescribed PhACs throughout the world are the psychiatric drugs that include the antidepressants and the antiepileptic drug carbamazepine (CAR) frequently used for treating schizophrenia and bi-polar disorder [7,8]. The persistent drug CAR largely sold in Canada is currently prescribed in combination to antidepressants all over the world during therapy. Therefore, a monitoring of CAR is also required to bet-ter assess its environmental fate in different matrices. Toxicity studies of these neuroactive compounds provided evidence for biological effects on aquatic organisms [9-13]. Although the occurrence of antidepressants in sewage effluents [6,14-17] and wastewater sludge [18-20] has been dem-onstrated, the fate of these substances following different treatments in STPs has not been extensively documented. A previous study indicated that a primary treatment process has limited capability to remove and/or degrade antidepressants residues in wastewater [15]. Further results ob-tained for STPs operating different biological processes (e.g. secondary treatment with activated sludge) revealed moderate potential (mean re-moval efficiency $\leq 30\%$) to degrade antidepressants from wastewater [20]. Therefore, alternative treatment technologies may have to be implemented or combined to achieve high removal of compounds in STPs [21]. As such, experimental evidence reported elsewhere clearly demonstrates that exist-ing limitations in primary and secondary processes can be overcome with more advanced treatment strategies including chemical oxidation with ozone or the use of high pressure membrane technologies [22-24].

While conventional activated sludge treatments were shown to degrade pharmaceuticals to varying extent [25], ozone (O_3) treatments showed promising results in terms of removal efficiencies as an efficient oxidizer to remove endocrine disruptors compounds and pharmaceuticals prod-ucts in wastewater [26,27]. Generally, O_3 reacts with organic molecules through either the direct reaction with molecular O_3 (via 1–3 dipolar cyclo addition reaction on unsaturated bonds, and electrophilic reaction on aro-matics having electron donor groups e.g. OH, NH_2) or by decomposition through the formation of chain intermediate free radicals, including the hydroxyl radical OH· (less selective reaction on saturated aliphatic mol-ecules) [26,28]. The stability of dissolved ozone is readily affected by pH, ultraviolet (UV) light, ozone concentration, and the concentration of radi-cal scavengers such carbonate—bicarbonate species, the dissolved organic carbon and humic acids [28,29]. Except for few experiments completed with fluoxetine (FLU), the number of studies dedicated to the elimination of antidepressants by oxidation processes (e.g. TiO_2 membrane reactor, O3 with UV activation, O_3 with H_2O_2) has been rather limited [22-24]. Since molecular O_3 is a selective electrophile that reacts quickly with amine and double bounds moieties [26], ozonation should be efficient to degrade antidepressants mostly constituted of secondary or tertiary amine and conjugated rings. However, as reported for β-Lactam antibacterial agents (e.g. penicillin G, cephalexin) spiked in wastewater, O_3 reaction leads to the formation of biologically active sulfoxides analogues [30]. For antidepressants, no study on the transformation products following an O_3 treatment in wastewater is currently available. As yet, no data is reported neither on by-products toxicity. Nevertheless, formation of N-oxide, amide, aldehyde, and carboxylic acid by-products is expected after ozonation of secondary and tertiary amine compounds in aqueous solutions [31,32].

In the present work, the effectiveness of ozone treatments in terms of removal efficiency is tested at three different concentrations for the oxidation of 14 antidepressants along with their direct N-desmethyl metabolites and the anticonvulsive drug carbamazepine during ozonation of a primary-treated effluent. The goal of the study was also to investigate the occurrence of antidepressant by-products formed in treated effluent after ozonation.

10.2 EXPERIMENTAL

10.2.1 CHEMICALS AND MATERIALS

All certified standards were > 98% purity grade. Fluoxetine (FLU), norfluoxetine (NFLU), paroxetine (PAR), sertraline (SER), (S)-citalopram (CIT), fluvoxamine (FLUVO), desmethylfluvoxamine (DFLUVO), mirtazapine (MIR), and desmethylmirtazepine (DMIR) were provided by Toronto Research Chemicals Inc. (North York, Ontario, Canada). Desmethylsertraline (DSER), venlafaxine (VEN), O-desmethylvenlafaxine (DVEN), and the surrogate standard bupropion-d_9 were obtained from Nanjing Jinglong PharmaTech (Nanjing, China). Amitriptyline (AMI), nortriptyline (NTRI), carbamazepine (CAR), and surrogate standard 10,11-dihydrocarbamazepine were purchased from Sigma-Aldrich Co.

(St. Louis, Missouri, USA), while internal standard cis-tramadol[13]C-d$_3$ was purchased from Cerilliant Corp. (Round Rock, Texas, USA). The high-performance liquid chromatography–grade solvents (methanol and aceto-nitrile) and ammonium hydroxide were provided by Caledon Laboratories Ltd. (Georgetown, Ontario, Canada). Reagent-grade hydrochloric acid, acetic acid, ammonium bicarbonate, and ACS grade ethyl acetate were provided by American Chemicals Ltd. (Montreal, Quebec, Canada). Solid-phase extraction (SPE) cartridges of 6 mL, 200 mg Strata™ X-C were pur-chased from Phenomenex (Torrance, California, USA). Stock solutions of 100 mg/L of each substance were prepared in methanol and stored at 4°C in amber glass bottles that were previously washed with methanol. The chemical structures of the selected compounds are provided in Figure 1.

10.2.2 INSTRUMENTATION

10.2.2.1 LIQUID CHROMATOGRAPHY (LC)

Liquid chromatography (LC) was performed using an Agilent 1200 Se-ries LC system equipped with binary pumps, degasser, and a thermostated autosampler maintained at 4°C. The antidepressants were separated on a Kinetex® XB-C18 column (100 mm × 2.10 mm, 1.7 µm) using a binary gradient made of (A) ammonium bicarbonate (5 mM) pH 7.8, and (B) acetonitrile at a flow rate of 400 µL/min. The volume of injection was 15 µL for influent, effluent, and sludge extracts. The gradient used was (%B): 0 min (10%), 6 min (80%), 10 min (80%), 12 min (90%), 14 min (10%), and 16 min (10%). An equilibration time of 4 min was used resulting in a total run time of 20 min. The column temperature was maintained at 40°C.

10.2.2.2 TANDEM-MASS SPECTROMETRY (QQQMS, QQTOFMS)

For quantitative analysis, the LC system was coupled to a 6410 triple quadrupole mass spectrometer (QqQMS) manufactured by Agilent Tech-nologies (Santa Clara, CA, USA) equipped with an electrospray ionization (ESI) source. The capillary was maintained at 4000 V, and the cone voltage was optimized for each compound in the positive-ion mode (ESI+). Additional detector parameters were held constant for all antidepressants: gas temperature 325°C; gas flow 10 L/min; nebulizer 35 psi and dwell time 50 ms. For qualitative by-products analysis, a 6530 quadrupole time-of-flight mass spectrometer (QqToFMS) also manufactured by Agilent Technologies, was utilized. The QqToFMS was equipped with a thermal gradient focusing ESI source (Jet Stream technology). Source parameters consisted of the following: gas temperature 325°C; sheath gas temperature 350°C; sheath gas flow 11 L/min; drying gas flow 5 L/min; nebulizer 35 psig, fragmentor 100 V and capillary voltage 4000 V. The QqToFMS was operated in the 4 GHz High Resolution mode with a low mass range (1700 m/z). Purine (121.050873 m/z) and Hexakis (922.009798 m/z) were used as internal reference masses to improve mass accuracy. Initial tests were performed on treated effluent extracts in high resolution tandem MS mode using a mass range of m/z 100–400 (specific collision energy: 0 V) at a rate of 5 spectra/s to screen the exact $[M+H]^+$ masses of the precursor ions. Identified compounds were then fragmented with different specific collision energies varying between 0 and 10 V. For both detection systems, the MassHunter software from Agilent Technologies was used for data acquisition and processing. Optimized parameters for QqQMS are listed in a table (Additional file 1).

10.2.3 SAMPLE LOCATION AND COLLECTION

10.2.3.1 SAMPLE LOCATION

All samples were collected onsite at the sewage treatment plant (STP) of the city of Repentigny (30 km North-East of Montreal, Qc, Canada) in amber glass bottles previously washed with methanol during an ozonation pilot-study performed in June 2011. The Repentigny STP typically treats 25 000 m³ of raw sewage daily for a population of approximately 60 000 persons. Wastewater is primary-treated using both physical and chemical treatments (e.g. flocculation of suspended matters with alum and/or $FeCl_3$). For the purpose of this study, treated wastewater was further experimentally ozone-oxidized on site. Main characteristics of the Repentigny STP are reported in Table 1. Ozonation of the effluent consisted of an ozone (O_3) generator (Ozone Solution, Model: TG10–Ozone Solution) fed with ultra-pure oxygen (99.9999%). Gaseous ozone was bubbled in a ceramic diffuser located inside a vertical column (6.3 m, 5.08 internal diameter) where both gas transfer and contact time occurred simultaneously. The water flow was maintained at 1.2 L/min, while the O3 flow rate injection was kept around 75 to 110 N mL/min (head pressure: 10 psi). Contact time of O_3 with treated effluent was 10 min. Ozone transfer was monitored by measuring off-gas ozone concentrations using the standard KI procedure [33]. Applied ozone dosages were then corrected for ozone transfer efficiency which varied from 75 to 80%. Total and residual dissolved O_3 concentrations were determined following the standard indigo trisulfonate colorimetric method [34].

TABLE 1: Main water characteristics of the Repentigny sewage treatment plant

Wastewater	Temperature (°C)	pH	Alkalinity (mg/L) $CaCO_3$	TSS (mg/L)	BOD$_5$ (mg/L)	COD (mg/L)
Raw sewage (Influent)	17	7.3	189	146	136	227
Effluent	–	7.2	165	12	36	59

TSS: Total Suspended Solids, BOD5: Biochemical Oxygen Demand, COD: Chemical Oxygen Demand.

10.2.3.2 SAMPLE COLLECTION

Typically, water samples of influent (raw sewage), primary-treated effluent, and ozone treated effluent were collected between 10:00 and 14:00 in polyethylene containers and stored on ice. Samples of wet primary sewage sludge (biosolids) were also collected on the same days and immediately stored on ice in polyethylene bottles. In the laboratory, approximately 10 g of wet biosolid material was filtered with a 0.7 μm glass fiber filter to get a dewatered sludge sample that was frozen, freeze-dried, and stored at −80°C until use. All samples were extracted and analyzed within 48 h after their collection.

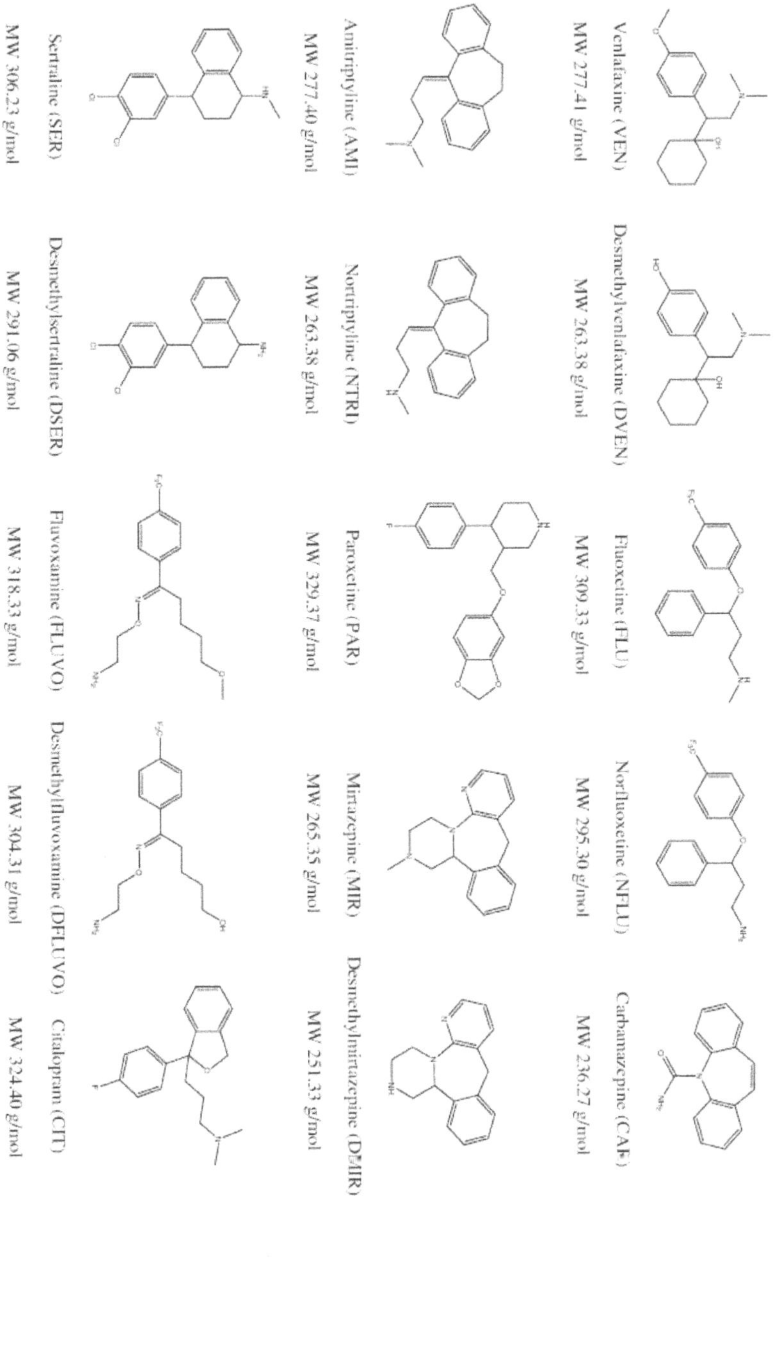

FIGURE 1: Chemical structures of the studied compounds.

Venlafaxine (VEN)
MW 277.41 g/mol

Desmethylvenlafaxine (DVEN)
MW 263.38 g/mol

Fluoxetine (FLU)
MW 309.33 g/mol

Norfluoxetine (NFLU)
MW 295.30 g/mol

Carbamazepine (CAF)
MW 236.27 g/mol

Amitriptyline (AMI)
MW 277.40 g/mol

Nortriptyline (NTRI)
MW 263.38 g/mol

Paroxetine (PAR)
MW 329.37 g/mol

Mirtazepine (MIR)
MW 265.35 g/mol

Desmethylmirtazepine (DMIR)
MW 251.33 g/mol

Sertraline (SER)
MW 306.23 g/mol

Desmethylsertraline (DSER)
MW 291.06 g/mol

Fluvoxamine (FLUVO)
MW 318.33 g/mol

Desmethylfluvoxamine (DFLUVO)
MW 304.31 g/mol

Citalopram (CIT)
MW 324.40 g/mol

10.2.4 SAMPLE EXTRACTION

10.2.4.1 SEWAGE SAMPLES

Extraction method for raw sewage and effluent samples to be analyzed for various classes of antidepressants was done as previously described [15]. The decision to incorporate the neutral drug carbamazepine (CAR) amongst the basic antidepressants forced us to modify the protocol by replacing the strong cation exchange cartridge by a mixed-mode cartridge for sample purification (Strata X-C, Phenomenex) [20].

The validated extraction protocol used here was similar to that described in Lajeunesse et al. [20]. Each 250 mL of filtered sewage sample were spiked with 100 µL of a surrogate standard solution prepared in methanol (bupropion-d9 / 10,11-dihydrocarbamazepine, 2.5 mg/L) and addition of 2.5 mL of methanol before lowering the pH to around 3 with 100 µL of phosphoric acid (85%). The mixed-mode solid phase extraction (SPE) cartridges were conditioned with 4 mL of methanol followed by at least 8 mL of Milli-Q water. SPE was performed with a VAC ELUT SPS24 manifold (Varian) at flow rates ~10–15 mL/min. After extraction, all cartridges were washed with 2 mL of HCl (0.1 M). The CAR molecules were eluted first with 2 × 2 mL of ethyl acetate prior the evaporation of the solvent in the tubes to dryness under a gentle stream of nitrogen. Meanwhile, all SPE cartridges were washed with 2 mL of methanol. The antidepressants retained onto the sorbent were then eluted with 2 × 2 mL of a solution of 5% (v/v) NH4OH in methanol. The combined fractions (e.g. CAR and antidepressants) were mixed with 100 µL of a solution of cis-tramadol[13]d$_3$ in methanol (5 mg/L) as the internal standard and the solvent in tubes was evaporated to dryness with nitrogen. The dried extracts were reconstituted with 0.50 mL of the mobile phase solution of ammonium bicarbonate (5 mM) pH 7.8 – acetonitrile (1:1 v/v) in injection vials and later injected in LC-QqQMS or LC-QqToFMS for analysis.

10.2.4.2 SEWAGE SLUDGE SAMPLES

The simultaneous extraction of CAR and antidepressants in biosolid samples was completed using the validated protocol reported in Lajeunesse et al. [20]. Briefly, 0.200 g of freeze-dried sludge is transferred to a 16 × 150 mm borosilicate glass screw-top conical tube before adding 8 mL of a solution composed of methanol / 0.1 M acetic acid buffer solution pH 4.0 (1:1 v/v). Each tube were spiked with 100 μL of a surrogate standard solution prepared in methanol (bupropion-d_9 / 10,11-dihydrocarbamazepine, 2.5 mg/L). Samples were then shaken vigorously and mixed on a rotary extractor (Caframo REAX) for 15 min. After extraction, tubes were placed in a sonication bath for 15 min before adding 4 mL of Milli-Q water to each tube. Tubes were then centrifuged (320 x g) at room temperature for 5 min. Following the SPE protocol described previously for aqueous sewage samples, supernatants were transferred directly on mixed-mode cartridges. The final extracts were reconstituted in 0.5 mL of the mobile phase solution of ammonium bicarbonate (5 mM) pH 7.8 – acetonitrile (1:1 v/v), filtered with a PTFE 0.45 μm filter, and then injected in LC-QqQMS system for analysis.

10.3 RESULTS AND DISCUSSION

10.3.1 ANTIDEPRESSANTS IN RAW SEWAGE AND PRIMARY-TREATED EFFLUENT

Out of the 15 compounds investigated, 13 were detected in raw sewage samples and only the antidepressant FLUVO and its direct metabolite DFLUVO were not detected. Compound concentrations ranged from 6.5 ng/L (NFLU) to 4185 ng/L (DVEN) (Table 2). A typical chromatogram of the detected antidepressants VEN, CIT, PAR, and FLU in a primary-treated effluent extract is depicted in Figure 2. Overall, moderate to poor removal efficiencies were obtained for most antidepressants (mean removal efficiency of 19%). Results showed that current enhanced primary treatment using physical and chemical processes removed little of the studied compounds (Table 2). The substances with lowest removal efficiencies were CAR (4.4%), along with the antidepressant metabolites NTRI (6.8%) and NFLU (7.1%). Similar low removal rates were previously reported for antidepressants [15] and CAR [35] in primary-treated effluents. Despite a noteworthy reduction of total suspended solids – TSS (Table 1), the weak removal obtained for this primary treatment strongly suggests that a mechanism other than chemical adsorption would be required to effectively remove antidepressants from urban wastewater.

TABLE 2: Mean concentrations of studied compounds extracted in wastewater (raw sewage, effluent) and biosolid samples from the Repentigny STP

Compounds	Wastewaters (n = 2)			Biosolids (n = 2)		
	Raw sewage (ng/L)	Effluent (ng/L)	Removal Eff. (%)	Sludge (ng/g)	K_d (L/kg)	log K_d
CIT	207 ± 12	148 ± 16	29	172 ± 38	1.2×10^3	3.1
SER	13 ± 1	9.4 ± 0.1	28	43 ± 5	4.6×10^3	3.7
DSER	23 ± 1	19 ± 3	17	31 ± 6	1.6×10^3	3.2
AMI	223 ± 21	195 ± 11	13	58 ± 22	2.9×10^2	2.5
NTRI	21 ± 3	19 ± 4	6.8	9.0 ± 1.1	4.7×10^2	2.7
VEN	4061 ± 153	3144 ± 107	23	227 ± 49	7.2×10^1	1.9
DVEN	4185 ± 133	3448 ± 279	18	73 ± 2	2.1×10^1	1.3
CAR	747 ± 14	714 ± 13	4.4	26 ± 12	3.6×10^1	1.6
FLU	11 ± 1	9.5 ± 0.6	16	15 ± 1	1.6×10^3	3.2
NFLU	7.0 ± 0.4	6.5 ± 0.2	7.1	3.8 ± 0.6	5.8×10^2	2.8
PAR	15 ± 1	13 ± 4	9.0	5.6 ± 3.6	4.2×10^2	2.6
MIR	171 ± 20	109 ± 3	36	27 ± 6	2.5×10^2	2.4
DMIR	41 ± 1	25 ± 1	38	13 ± 1	5.4×10^2	2.7

10.3.2 ANTIDEPRESSANTS IN SEWAGE SLUDGE

Primary sludge samples consistently displayed quantifiable amounts of the studied compounds (excepted FLUVO and DFLUVO) (Table 2). Highest mean concentrations in biosolid samples were found for VEN (227 ng/g), CIT (172 ng/g), DVEN (70 ng/g), AMI (58 ng/g), and SER (43 ng/g). Our results are consistent with the mean concentrations for the antidepressants FLU (123 ng/g) and PAR (41 ng/g) reported by Radjenović et al. [18] in primary sludge samples. Interestingly, among reported concentrations, less antidepressant metabolites were detected in sewage sludge samples for N-desmethyl metabolites in comparison to their respective parent molecules. These findings suggest that more polar compounds have a lower affinity for the solid phase of sewage sludge and hence have limited removal efficiencies.

TABLE 3: Mean concentrations and removal of antidepressants contained in final effluent following ozonation

Compounds	Conc. (n=2) Ozone 5 mg/L			Conc. (n=2) Ozone 9 mg/L		
	Effluent (ng/L)	Disinfected effluent (ng/L)	Removal Eff. (%)	Effluent (ng/L)	Disinfected effluent (ng/L)	Removal Eff. (%)
CIT	186 ± 27	123 ± 11	34	148 ± 16	56 ± 1	62
SER	14 ± 2	–	100	9.4 ± 0.1	–	100
DSER	23 ± 1	–	100	19 ± 3	–	100
AMI	106 ± 5	36 ± 1	66	195 ± 11	15 ± 1	92
NTRI	18 ± 1	0.18 ± 0.01	99	19 ± 4	–	100
VEN	2194 ± 191	963 ± 43	56	3144 ± 107	986 ± 27	69
DVEN	2319 ± 11	–	100	3448 ± 279	–	100
CAR	716 ± 4	12 ± 1	98	714 ± 13	–	100
FLU	6.3 ± 0.8	–	100	9.5 ± 0.6	–	100
NFLU	11 ± 2	–	100	6.5 ± 0.2	–	100
PAR	9.0 ± 1.3	–	100	13 ± 4	–	100
MIR	104 ± 1	1.6 ± 0.1	98	109 ± 3	–	100
DMIR	41 ± 4.1	3.3 ± 0.4	92	25 ± 1	–	100

Note: Measured residual O_3 concentrations for 5, 9 and 13 mg/L of O_3 were respectively 0.000, 0.036 and 0.514 mg/L.

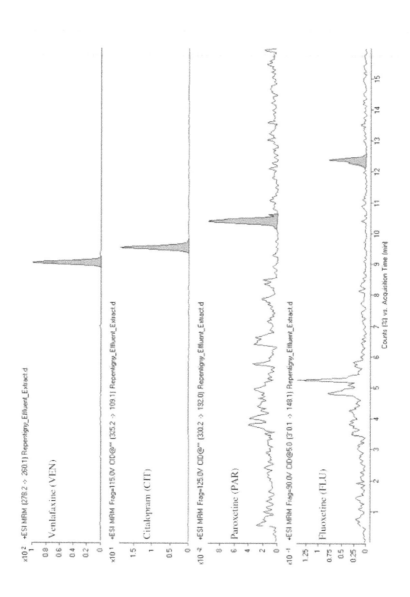

FIGURE 2: Representative LC-QqQMS chromatograms of selected antidepressants detected in primary-treated effluent sample extract.

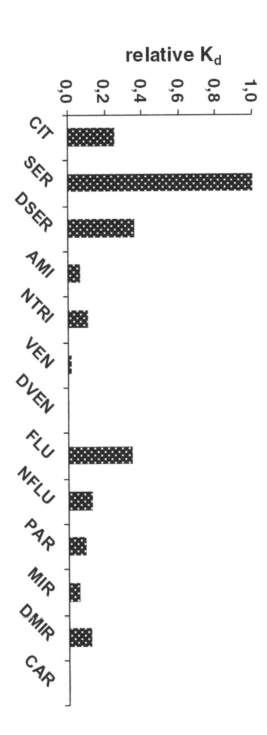

FIGURE 3: Relative K_d values of the studied compounds.

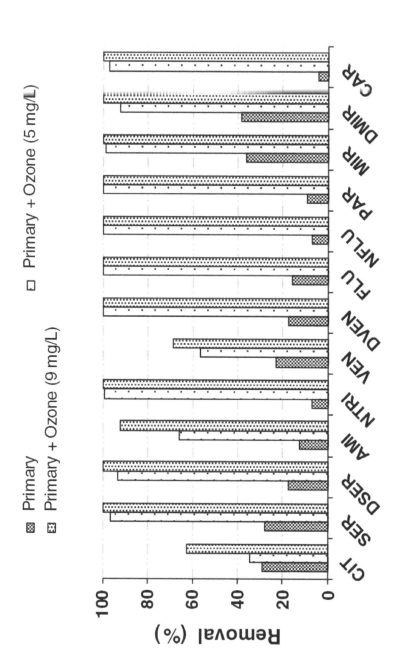

FIGURE 4: Reduction of antidepressants and CAR in primary-treated effluent by ozone disinfection at 5 and 9 mg/L O_3 concentration.

In order to describe the fate and behavior of antidepressants in primary STP, specific partitioning coefficient (K_d) values for antidepressants and metabolites to sewage sludge were estimated. The K_d coefficients were calculated using the ratio [Sludge] / [Effluent]; where [Sludge] is the concentration of antidepressants in sewage sludge (ng/kg) and [Effluent] is the concentrations of antidepressants in final effluent (ng/L) [36]. The obtained Kd values were applied to evaluate the affinity of compounds to primary STP sludge. The Kd values were lowest for VEN, DVEN, and CAR (Table 2) with values ranging from 21 to 72 L/kg. With log Kd values ≤ 2, sorption to solid matter for VEN, DVEN, and CAR is therefore defined as negligible [36]. Higher sorption behaviour is expected for SER, DSER, FLU, and CIT which have higher relative K_d values (Figure 3).

10.3.3 ANTIDEPRESSANTS IN TREATED EFFLUENT—OZONATION

Ozonation of the primary-treated effluent did degrade antidepressants with higher efficiency, yielding a mean removal efficiency of 88% when 5 mg/L of ozone was applied (Table 3). Ten (10) of the 13 compounds initially present in the effluent had removal efficiencies ≥ 92% (Figure 4). Only three substances (CIT, AMI, and VEN) yielded lower removal efficiencies, being 34, 66, and 56% respectively. As discussed in background section, the ozonation mechanism is directly affected by the ozone stability. Thus, scavengers compounds (e.g. carbonate, bicarbonate, dissolved organic and humic acids) present in effluent may have slowed down the ozone decomposition by inhibiting the free-radical reaction chain, and consequently the formation of hydroxyl radicals OH· necessary to degrade saturated aliphatic carbon chain on molecules [28]. Since, CIT, AMI and VEN have long tertiary amine aliphatic chains on their chemical structures, steric hindrance may have prevented ozone reactions normally expected at specific sites of the molecules [37]. In present study, it is very difficult to assess the relative importance of direct ozone-mediated transformations, and thereby to draw a general conclusion about each compound and transformation during ozonation in a single matrix with varying OH· scavenging capacities, under a certain pH condition. Obviously, the work presented therein was not intended to the understanding of ozonation mechanisms. However, as reported by Zwiener and Frimmel [38], so-called radical scavengers compete with pharmaceuticals for the OH-radicals and by this decrease the degradation kinetics of the targeted pharmaceuticals. Nevertheless, removal efficiency increased to 94% for most compounds using an optimal ozone dose of 9 mg/L (Figure 4). At the highest ozone treatment tested (i.e. 13 mg/L), all antidepressants were oxidized and degraded from primary-treated effluent samples. Current limitation of the analytical method may have lead to undetected polar compounds that would require different chromatographic and instrumental adjustments. However, Snyder et al. [26] have reported very similar removal efficiencies for CAR (> 99%) and FLU (> 93%) for comparable effluent samples treated with 3.6 mg/L of O_3. Under controlled conditions using a 5–L glass jacketed reactor, Rosal et al. [39] observed high removal efficiencies for CAR (98%), CIT (93%), FLU (100%), and VEN (88%) in wastewater samples exposed to 2.4 – 6.1 mg/L of O_3 for less than 5 min.

10.3.4 CHARACTERIZATION OF NEW BY-PRODUCTS BY LC-QQTOFMS

In this study, the two most abundant antidepressants detected in raw sewage were VEN and its N-desmethyl metabolite DVEN. Therefore, primary-treated effluent samples previously treated with O_3 at different concentrations were screened by LC-QqToFMS to confirm the presence of related by-products of these two compounds.

Initial tests performed on treated effluent extracts (O_3 dose: 5 mg/L) in high resolution tandem MS mode using a mass range of m/z 100–400 (specific collision energy: 0 V) enabled the positive detection of N-oxide by-product precursor ions for VEN (m/z 294.2059, accurate mass error: –3.40 ppm) and DVEN (m/z 280.1903, accurate mass error: –3.21 ppm). The chromatograms and mass spectrums of both characterized by-products are depicted respectively in Figures 5, 6a, and 7a. Precursor $[M + H]^+$ ions were isolated in the first quadrupole of the QqToF and then fragmented in the collision cell at 10 V in order to perform accurate mass measurements on the resulting fragment ions. Isolation and fragmentation of the precursor ion of N-oxide VEN (m/z 294.2057, accurate mass error: -4.08 ppm) generated a product ion at m/z 127.1125 (Figure 6b). This ion fragment corresponds to $[C_8H_{14}O + H]^+$ and has an accurate mass error from theoretical values of 1.57 ppm. As for the N-oxide DVEN when its precursor ion at m/z 280.1910 (accurate mass error: −0.71 ppm) was isolated and fragmented, an ion at m/z 113.0966 was observed that could be interpreted as $[C_7H_{12}O + H]^+$ with an accurate mass error of ± 0.00 ppm (Figure 7b). During MS/MS characterization, it was decided to keep a large isolation width of the quadrupole (e.g. 4 m/z) to increase sensitivity. Hence, MS/MS mass spectra of N-oxide VEN and DVEN likely contained product ions of other molecules that may have interfered with the mass spectra interpretation. According to European Commission Decision 2002/657/EC [40], at least 4 "identification" points are required in order to confirm the presence of a substance. Since one high-resolution precursor ion and one high-resolution product ion were obtained during experiments (total identification points: 2 + 2.5 = 4.5), the results of our study (with accurate mass errors < ± 5.00 ppm) were considered sufficient to confirm the presence of the N-oxide by-products.

Additional LC-QqToFMS analysis performed on effluent extracts previously treated with 9 mg/L of O_3 confirmed also the presence of both N-oxide by-products. When the concentration of O_3 reached 13 mg/L, none of the by-products were detected in corresponding effluent samples. This suggests that an optimal O_3 dosage would be required to completely degrade the N-oxide by-products from treated effluents. Additional tests performed on raw sewage (influent) and primary-treated effluent confirmed the absence of the two N-oxide by-products prior ozone treatments. To our knowledge, the present study is the first one to report the characterization of antidepressant by-products in municipal effluent samples after experimental ozone treatment.

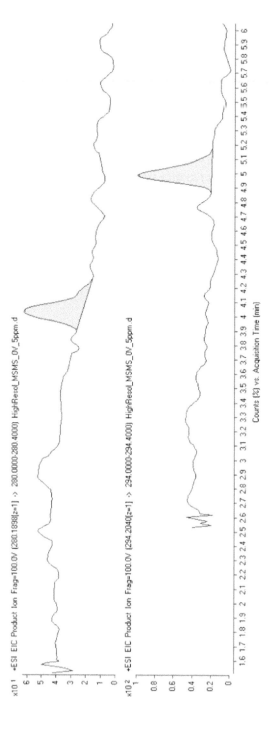

FIGURE 5: LC-QqToFMS chromatograms of N-oxide by-products detected in disinfected effluent (O3 concentration: 5 mg/L).

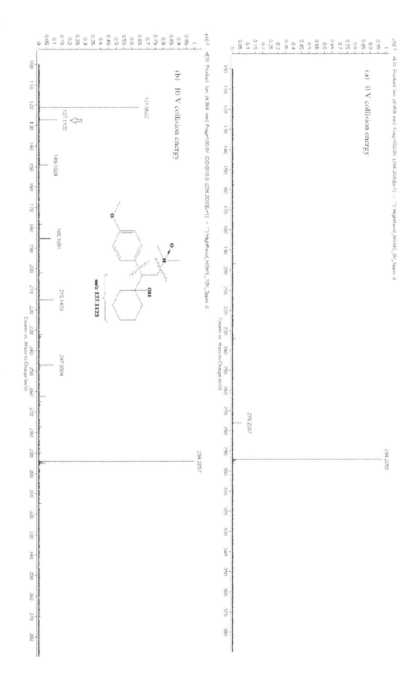

FIGURE 6: LC-QqToFMS mass spectra of N-oxide VEN in disinfected effluent (O3 concentration: 5 mg/L): product ions at 0 V collision energy (a) and 10 V collision energy (b).

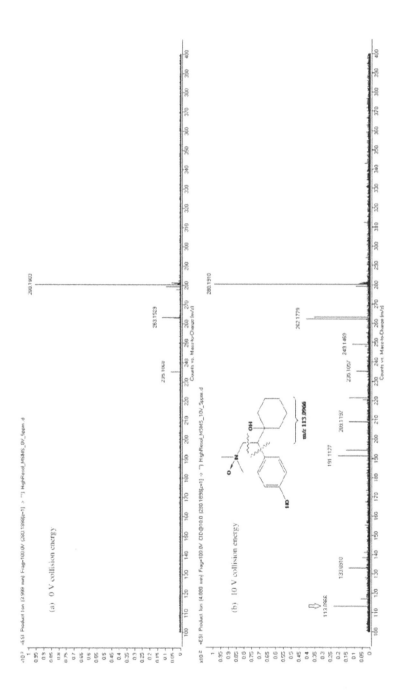

FIGURE 7: LC-QqToFMS mass spectra of N-oxide DVEN in disinfected effluent (O3 concentration: 5 mg/L): product ions at 0 V collision energy (a) and 10 V collision energy (b).

10.4 CONCLUSIONS

This study described the fate and behavior of antidepressants and their N-desmethyl metabolites in a primary STP following ozone treatment. Effluent ozonation led to higher mean removal efficiencies than current primary treatment, and therefore has represented a promising strategy for the elimination of antidepressants in urban wastewaters. However, the use of O_3 has produced N-oxide by-products with unknown toxicity. Of particular concern is the potential that removal of pharmaceuticals following wastewater disinfection using advanced oxidation process (i.e. ozonation) could generate by-products of similar parent chemical structures that would need to be identified, quantified and evaluated for their toxicity.

REFERENCES

1. Halling-Sørensen B, Nors Nielsen S, Lanzky PF, Ingerslev F, Holten Lützhøft HC, Jørgensen SE: Occurence, fate and effect of pharmaceutical substances in the environment – A review. Chemosphere 1998, 36:357-393.
2. Daughton EG, Ternes TA: Pharmaceuticals and personal care products in the environment: agents of subtle change? Environ Health Perspect 1999, 107:907-938.
3. Heberer T: Occurence, fate and removal of pharmaceuticals residues in the aquatic environment: A review of recent research data. Toxicol Lett 2002, 131(1-2):5-17.
4. Ternes TA: Occurrence of drugs in German sewage plants and rivers. Water Res 1998, 32:3245-3260.
5. Oulton RL, Kohn T, Cwiertny DM: Pharmaceuticals and personal care products in effluent matrices: a survey of transformation and removal during wastewater treatment and implications for wastewater management. J Environ Monit 2010, 12:1956-1978.
6. Vasskog T, Anderssen T, Pedersen-Bjergaard S, Kallenborn R, Jensen E: Occurrence of selective serotonin reuptake inhibitors in sewage and receiving waters at Spitsbergen and in Norway. J Chromatogr A 2008, 1185:194-205.
7. Van Rooyen GF, Badenhorst D, Swart KJ, Hundt HKL, Scanes T, Hundt AF: Determination of carbamazepine and carbamazepine 10,11-epoxide in human plasma by tandem liquid chromatography-mass spectrometry with electrospray ionisation. J Chromatogr B 2002, 769:1-7.
8. Calisto V, Esteves VI: Psychiatric pharmaceuticals in the environment. Chemosphere 2009, 77:1257-1274.
9. Fong PP: Antidepressants in aquatic organisms: a wide range of effects. In Pharmaceuticals and Personal Care Products in the environment. Edited by Daughton CG,

Jones-Lepp TL. Washington, USA: Scientific and regulatory issue, ACS Symposium series; 2001:264-281.

10. Gagné F, Blaise C, Fournier M, Hansen PD: Effects of selected pharmaceutical products on phagocytic activity in Elliptio complanata mussels. Comp Biochem Physiol 2006, C143:179-186.

11. Mennigen JA, Lado WE, Zamora JM, Duarte-Guterman P, Langlois VS, Metcalfe CD, Chang JP, Moon TW, Trudeau VL: Waterborne fluoxetine disrupts the reproductive axis in sexually mature male goldfish, Carassius auratus. Aquat Toxicol 2010, 100:354-364.

12. Lajeunesse A, Gagnon C, Gagné F, Louis S, Čejka P, Sauvé S: Distribution of antidepressants and their metabolites in brook trout exposed to municipal wastewaters before and after ozone treatment – Evidence of biological effects. Chemosphere 2011, 83:564-571.

13. Lazzara R, Blázquez M, Porte C, Barata C: Low environmental levels of fluoxetine induce spawning and changes in endogenous estradiol levels in the zebra mussel Dreissena polymorpha. Aquat Toxicol 2012, 106–107:123-130.

14. Rúa-Gómez P, Püttmann W: Impact of wastewater treatment plant discharge of lidocaine, tramadol, venlafaxine and their metabolites on the quality of surface waters and groundwater. J Environ Monit 2012, 14:1391-1399.

15. Lajeunesse A, Gagnon C, Sauvé S: Determination of basic antidepressants and their N-desmethyl metabolites in raw sewage and wastewater using solid-phase extraction and liquid chromatography-tandem mass spectrometry. Anal Chem 2008, 80:5325-5333.

16. Schultz MM, Furlong ET: Trace analysis of antidepressants pharmaceuticals and their select degradates in aquatic matrixes by LC/ESI/MS/MS. Anal Chem 2008, 80:1756-1762.

17. Metcalfe CD, Chu S, Judt C, Li H, Oakes KD, Servos MR, Andrews DM: Antidepressants and their metabolites in municipal wastewater, and downstream exposure in an urban watershed. Environ Toxicol Chem 2010, 29:79-89.

18. Radjenović J, Jelić A, Petrović M, Barceló D: Determination of pharmaceuticals in sewage sludge by pressurized liquid extraction (PLE) coupled to liquid chromatography-tandem mass spectrometry. Anal Bioanal Chem 2009, 393:1685-1695.

19. Hörsing M, Ledin A, Grabic R, Fick J, Tysklind M, la Cour Jansen J, Andersen HR: Determination of sorption of seventy-five pharmaceuticals in sewage sludge. Water Res 2011, 45:4470-4482.

20. Lajeunesse A, Smyth SA, Barclay K, Sauvé S, Gagnon C: Distribution of antidepressant residues in wastewater and biosolids following different treatment processes by municipal wastewater treatment plants in Canada. Water Res 2012, 46:5600-5612.

21. Oller I, Malato S, Sánchez-Pérez JA: Combination of advanced oxidation Processes and biological treatments for wastewater decontamination – A review. Sci Tot Environ 2011, 409:4141-4166.

22. Benotti M, Stanford B, Wert E, Snyder S: Evaluation of a photocatalytic reactor membrane pilot system for the removal of pharmaceuticals and endocrine disrupting compounds from water. Water Res 2009, 43:1513-1522.

23. Wert E, Rosario-Ortiz F, Snyder S: Effect of ozone exposure on the oxidation of trace organic contaminants in wastewater. Water Res 2009, 43:1005-1014.

24. Méndez-Arriaga F, Otsu T, Oyama T, Gimenez J, Esplugas S, Hidaka H, Serpone N: Photooxidation of the antidepressant drug fluoxetine (Prozac®) in aqueous media by hybrid catalytic/ozonation processes. Water Res 2011, 45:2782-2794.

25. Huber MM, Göbel A, Joss A, Hermann N, Löffler D, Mcardell CS, Ried A, Siegrist H, Ternes TA, von Gunten U: Oxidation of pharmaceuticals during ozonation of municipal wastewater effluent: A pilot study. Environ Sci Technol 2005, 39:4290-4299.

26. Snyder SA, Wert EC, Rexing DJ, Zegers RE, Drury DD: Ozone oxidation of endocrine disruptors and pharmaceuticals in surface water and wastewater. Ozone Sci Eng 2006, 28:455-460.

27. Gagnon C, Lajeunesse A, Cejka P, Gagné F, Hausler R: Degradation of selected acidic and neutral pharmaceutical products in a primary-treated wastewater by disinfection processes. Ozone Sci Eng 2008, 30:387-392.

28. Langlais B, Reckhow DA, Brink DR: Chapter II: Fundamental aspects. In Ozone in water treatment – Application and engineering. AWWA Research Association / Compagnie Générale des eaux. Edited by Langlais B, Reckhow DA, Brink DR. Michigan, USA: Lewis publishers inc; 1991:11-79.

29. Tomiyasu H, Fukutomi H, Gordon G: Kinetics and mechanism of ozone decomposition in basic aqueous solution. Inorg Chem 1985, 24:2962-2966.

30. Dodd MC, Rentsch D, Singer HP, Kohler H-PE, von Gunten U: Transformation of β-Lactam antibacterial agents during aqueous ozonation: reaction pathways and quantitative bioassay of biologically-active oxidation products. Environ Sci Technol 2010, 44:5940-5948.

31. Elmghari-Tabib M, Laplanche A, Venien F, Martin G: Ozonation of amines in aqueous solutions. Water Res 1982, 16:223-229.

32. Elmghari-Tabib M, Dalouche A, Faujour C, Venien E, Martin G, Legeron JP: Ozonation reaction patterns of alcohols and aliphatic amines. Ozone Sci Eng 1982, 4:195-205.

33. Standard method 2350. Oxidant demand / Requirement. Approved by SM Committee; 2007.

34. U.S. Environmental Protection Agency (EPA): Standard method 4500–03 B for ozone. 21st edition. U.S: Environmental Protection Agency (EPA); 2005. [Standard methods for the examination of water and wastewater]

35. Lajeunesse A, Gagnon C: Determination of acidic pharmaceutical products and carbamazepine in roughly primary-treated wastewater by solid-phase extraction and gas chromatography-tandem mass spectrometry. Intern J Environ Anal Chem 2007, 87:565-578.

36. Deegan AM, Shaik B, Nolan K, Urell K, Oelgemöller M, Tobin J, Morrisey A: Treatment options for wastewater effluents from pharmaceutical companies. Int J Environ Sci Tech 2011, 8:649-666.

37. Trimm DL: Chapter 4: The liquid phase oxidation of sulphur, nitrogen, and chlorine compounds. In Comprehensive chemical kinetics – Vol. 16 Liquid phase oxidation. Edited by Bamford CH, Tipper CFH. Amsterdam, The Netherlands: Elsevier Scientific Publishing Company; 1980:205-249.

38. Zwiener C, Frimmel FH: Oxidative treatment of pharmaceuticals in water. Water Res 2000, 34(6):1881-1885.

39. Rosal R, Rodríguez A, Perdigón-Melón JA, Petre A, García-Calvo E, Gómez MJ, Agüera A, Fernández-Alba AR: Occurrence of emerging pollutants in urban wastewater and their removal through biological treatment followed by ozonation. Water Res 2010, 44:578-588.

40. Commission of the European Communities: Commission Decision (2002/657/EC) of 12 August 2002: Implementing Council Directive 96/23/EC concerning the performance of analytical methods and the interpretation of results. Off J Eur Commun 2002, (8):L221-17. Internet access: http://www.ecolex.org.

PART IV

TEXTILE INDUSTRY

CHAPTER 11

Removal of Textile Dyes from Aqueous Solution by Heterogeneous Photo-Fenton Reaction Using Modified PAN Nanofiber Fe Complex as Catalyst

XUETING ZHAO, YONGCHUN DONG, BOWEN CHENG, AND WEIMIN KANG

11.1 INTRODUCTION

The dyeing wastewaters containing remarkable concentration of organic dyes are often produced by the textile industry around the world. It is well known that these dyes are toxic and nonbiodegradable to aquatic animals and plant [1–4]. Therefore, the treatment of dyeing wastewaters has become a research focus in the field of environment research in recent years. Various physically or mechanically based techniques (e.g., activated carbon adsorption, coagulation/flocculation, or membrane separation) have been extensively utilized. However, they only do a phase transfer of the pollutants. Compared with these techniques, Fenton and photo-Fenton processes as the advanced oxidation technologies can potentially complete the mineralization of organic pollutants in water. Moreover, the heterogeneous Fenton technology is replacing the homogeneous system be-

Removal of Textile Dyes from Aqueous Solution by Heterogeneous Photo-Fenton Reaction Using Modified PAN Nanofiber Fe Complex as Catalyst.© Zhao X, Dong Y, Cheng B, and Kang W. International Journal of Photoenergy **2013** (2013), http://dx.doi.org/10.1155/2013/820165. *Licensed under a Creative Commons Attribution 3.0 Unported License, http://creativecommons.org/licenses/by/3.0/.*

cause of its special advantages on a wide range of pH adaptability and easy recycling use [5]. The heterogeneous Fenton catalysts can be obtained by immobilizing Fe ions on the polymer substrates such as Nafion membrane and fibrous materials. During the past decade years, we have been devoted to exploring the polyacrylonitrile (PAN) fiber supported Fenton catalysts. And the resulting modified PAN fiber Fe complexes have been considered as the effective heterogeneous Fenton catalysts with low cost [6]. But this catalyst suffers from some problems, such as low specific surface area and low utilization stability. Therefore, it is necessary to explore new heterogeneous catalysts with higher performance for the dye degradation using the polymer materials with larger surface area as the support substrates. On the other hand, PAN nanofiber is regarded as one of the attractive fibrous materials owing to its excellent characteristics including large surface area, remarkably high porosity, and high permeability compared with conventional fibers [7, 8]. It has been applied widely in fine filtration [9, 10], enzyme immobilization [11–14], and adsorptive membranes [15–18]. However, the detailed information regarding PAN nanofiber as the supported material for heterogeneous Fenton catalyst is very limited. In the present work, PAN nanofibers were firstly prepared by the electrospinning technique. PAN nanofiber Fe complex was then produced by its surface modification and metal coordination. Moreover, the activity of this complex was investigated as a heterogeneous Fenton catalyst for the oxidative degradation of the dye with H_2O_2. Finally, some factors affecting the dye degradation such as Fe content of catalyst, light irradiation, H_2O_2 initial concentration as well as structure and initial concentration of the dyes used were optimized and discussed.

11.2 EXPERIMENTAL

11.2.1 MATERIALS AND REAGENTS

PAN knitting bulky yarn (PAN yarn) that consisted of twisted PAN fibers was obtained from Shanghai Shilin Spinning Company. N,N-dimethylformamide (DMF), hydroxylamine hydrochloride, hydrogen peroxide (30%), and ferric chloride were of laboratory agent grade. Three kinds of textile dyes including Reactive Red 195 (RR 195), Reactive Blue 4 (RB 4), and Acid Blue 7 (AB 7) were commercially available and purified by reprecipitation method in this experiment. Their molecular structures are presented in Figure 1. Double distilled and deionized water was used throughout the study.

11.2.2 ELECTROSPINNING OF PAN NANOFIBER

According to [11, 19], the 14 wt% solution of PAN was prepared by dissolving PAN yarn in DMF under constant stirring at room temperature for 24 h. The obtained PAN solution was added to a 10 mL glass syringe with a blunt needle. The PAN solution flow rate was controlled by a microinfusion pump to be 0.30 mlh^{-1}. The high-voltage supplier was used to connect the grounded collector and metal needles for forming electrostatic fields. The used voltage was 20 kV and a piece of flat aluminum foil on the grounded collection roller was placed about 15 cm below the tip of the needle to collect the nanofiber. A jet of PAN solution came out from the needle tip at a critical voltage and was collected on the aluminum foil. The PAN nanofibers were obtained after DMF evaporation.

11.2.3 PREPARATION OF MODIFIED PAN NANOFIBER FE COMPLEX

On the basis of our previous work [20], PAN nanofibers were amidoximated using a mixed solution containing hydroxylamine hydrochloride and sodium hydroxide in a 250 mL flask with a thermometer and agitator for at 1-2 h at 68°C. And then the resulting amidoximated PAN nanofiber (denoted as AO-n-PAN) was washed several times with distilled water and dried under vacuum at 50°C. The degree of the conversion from nitrile groups to amidoxime groups of the AO-n-PAN was calculated to be 41.21% [19]. The obtained AO-n-PAN nanofiber was immersed in an aqueous solution of ferric chloride under continuous agitation at 50°C for 2 h in order to produce the amidoximated PAN nanofiber Fe complex (denoted as Fe-AO-n-PAN). The residual concentration of Fe^{3+} ions in solution after coordination was determined using a VISTA MPX inductively coupled plasma optical emission spectrometer (ICP-OES, Varian Corp., USA) for calculating the Fe content (Q_{FE}) of Fe-AO-n-PAN.

11.2.4 CHARACTERIZATION OF MODIFIED PAN NANOFIBER FE COMPLEX

The surface morphology of Fe-AO-n-PAN was examined by a field emission scanning electron microscope (FE-SEM) (S-4800, Hitachi Co., Japan). The average diameter of PAN nanofibers was determined by analyzing the SEM images with an image analyzing software (Image-Pro Plus, Media Cybernetics Inc.) [21]. Composition of Fe-AO-n-PAN was verified by using a Nicolet Magana-560 Fourier transform spectrometer (Nicolet Instrument Co., USA).

11.2.5 PHOTOCATALYTIC REACTION SETUP

The photoreaction system was particularly designed in this experiment and consisted primarily of chamber, lamp, electromagnetic valve, relay, and water bath [22]. Ten open Pyrex vessels of 150 mL capacity were served as reception receivers in water bath. A 400 W high-pressure mercury lamp was used as illuminating source for the photocatalytic reaction. The intensity of light irradiation over the surface of the test solution was measured to be 8.42 mWcm^{-2} (400–1000 nm) and 0.62 mWcm^{-2} (UV 365 nm) using FZ-A radiometer and UV-A radiometer (Beijing BNU Light and Electronic Instrumental Co., China), unless otherwise stated.

11.2.6 DYE DEGRADATION PROCEDURE AND ANALYSIS

0.20 g of Fe-AO-n-PAN was immersed into the 50 mL test solutions containing 0.05 mmol L^{-1} dye and the appropriate concentration of H_2O_2 in the vessel. The temperature in the vessel was kept at 25 ± 1°C. The solution in vessel was exposed to the irradiation of lamp in photoreaction system. At the irradiation time intervals, 1-2 mL of the test solution was taken off from the vessel and rapidly measured at the λ_{max} of the dye used (522 nm for RR 195, 596 nm for RB 4, and 630 nm for AB 7) using a UV-2401 Shimadzu spectrophotometer. The decoloration percentage of the dye was expressed as D% = (1 – C/C$_0$) x 100%, where C$_0$ is the initial concentration of the dye (mmol L^{-1}) and C is the residual concentration of the dye (mmol L^{-1}). In addition, total organic carbon (TOC) of the test solution was measured by a Phoenix 8000 TOC analyzer (Tekmar-Dohrmann Inc., USA), and TOC removal percentage of the test solution was calculated as follows: TOC removal% = (1 – TOC$_t$/TOC$_0$) x 100%, where TOC$_0$ and TOC$_t$ are the TOC values (mgL^{-1}) at reaction times 0 and t, respectively.

11.3 RESULTS AND DISCUSSION

11.3.1 MORPHOLOGICAL PROPERTIES AND STRUCTURE

Figure 2 shows the random orientation and smooth surface of PAN nanofiber, and their average diameters are 319.8 nm. Moreover, the surface of PAN nanofibers is rarely affected by the modification and Fe coordination. However, the average diameters of the resulting Fe-AO-n-PAN are 512.8 nm.

Figure 3(a) exhibited the FTIR spectrum of PAN nanofiber. The typical characteristic bands due to the stretching vibration of hydroxyl (3540 cm^{-1}), nitrile (2248 cm^{-1}), carbonyl (1738 cm^{-1}), and ether (1250 and 1150–1040 cm^{-1}) were clearly observed [6, 16]. The intensity of CN group decreased after modification with hydroxylamine. The characteristic bands of the functional groups of PAN with additional peaks appeared at 1662 cm^{-1} and 944 cm^{-1} which is due to the stretching vibration of N–O, N–H and C=N groups in amidoxime, respectively [16]. The intensity of these peaks further decreased and two peaks at 1662 and 944 cm^{-1} shifted to short position, separately after coordinating with Fe^{3+} ions in the spectrum of Fe-AO-n-PAN. This demonstrated that the Fe^{3+} ions were chemically attached to the AO-n-PAN.

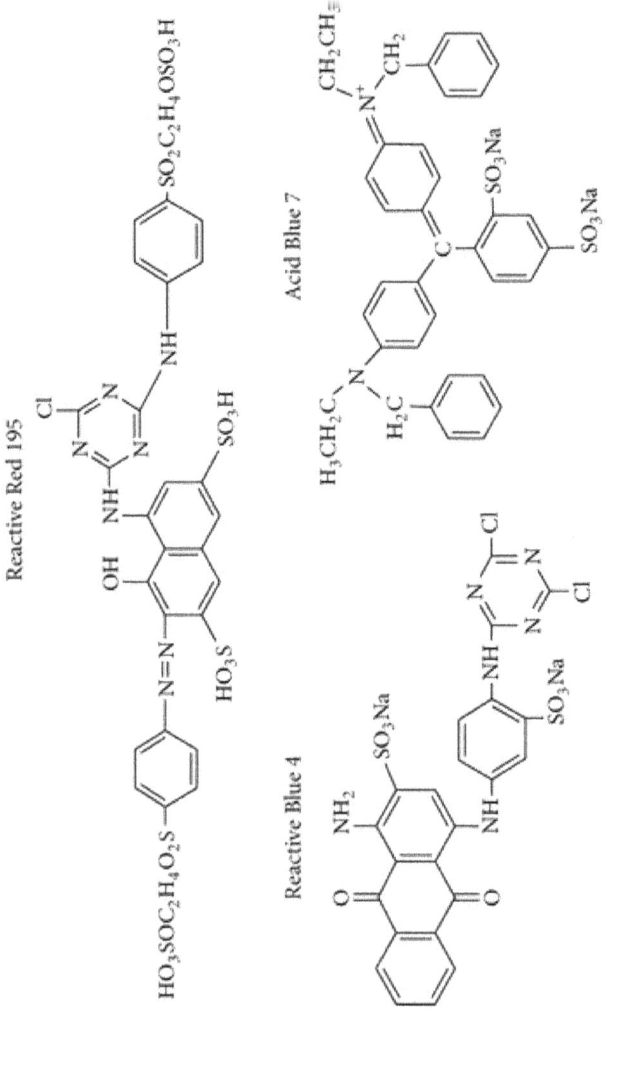

FIGURE 1: Chemical structures of the three dyes.

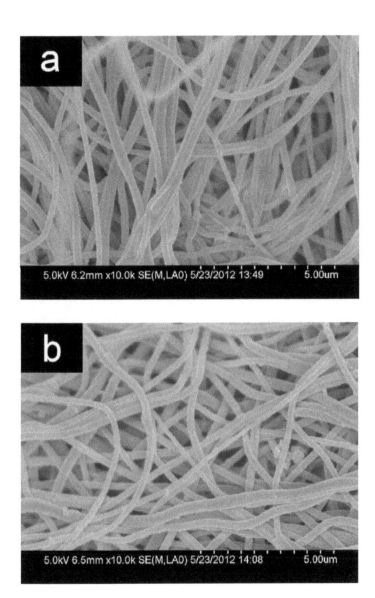

FIGURE 2: SEM images of (a) PAN nanofibers and (b) Fe-AO-n-PAN.

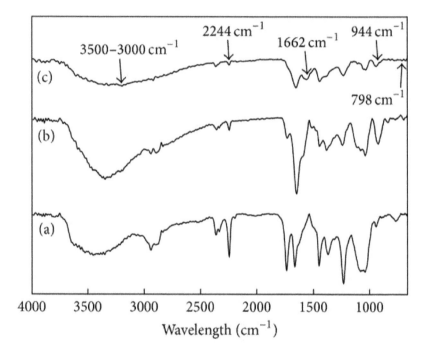

FIGURE 3: The FTIR spectra of (a) PAN nanofiber, (b) AO-n-PAN, and (c) Fe-AO-n-PAN.

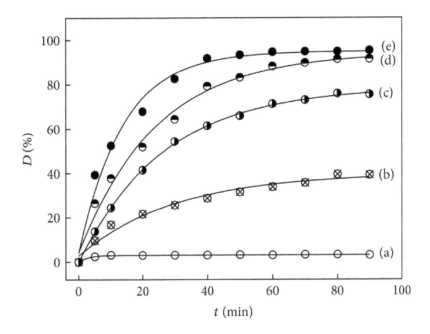

FIGURE 4: Variations in D% values of RR 195 in different systems: (a) RR 195 and H_2O_2 under visible irradiation, (b) RR 195, Fe-AO-PAN and H_2O_2 under dark, (c) RR 195, Fe-AO-PAN, and H_2O_2 under visible irradiation, (d) RR 195, Fe-AO-n-PAN, and H_2O_2 under dark, and (e) RR 195, Fe-AO-n-PAN, and H_2O_2 under visible irradiation. The experiments were carried out under conditions: $[H_2O_2] = 3.0\,mmol\,L^{-1}$, pH = 6, Fe-AO-PAN ($Q_{FE} = 3.42\,mmol\,g^{-1}$), or Fe-AO-n-PAN ($Q_{FE} = 3.38\,mmol\,g^{-1}$) = 0.20 g.

11.3.2 CATALYTIC ACTIVITY OF FE-AO-N-PAN

To study the catalytic property of Fe-AO-n-PAN as a heterogeneous Fenton photocatalyst, the degradation of RR 195 was conducted under the various conditions. For comparison, a control experiment was also performed in the presence of the amidoximated PAN fiber Fe complex (denoted as Fe-AO-PAN) with the PAN yarns using the same method described above, and the results were given in Figure 4.

Figure 4 shows that the D% value is less than 4.0% without Fe-AO-n-PAN or Fe-AO-PAN after 90 min of the reaction time under visible irradiation, demonstrating that the decomposition of RR 195 is rather slow in the absence of the catalyst. It is clearly found that when Fe-AO-n-PAN or Fe-AO-PAN was used, the D% value increased significantly with the prolongation of the reaction time. These results suggest that Fe-AO-n-PAN or Fe-AO-PAN exhibits strong catalytic activity on the dye degradation. This is because the Fe-AO-n-PAN or Fe-AO-PAN can enhance the reduction of the loaded Fe^{3+} to Fe^{2+} ions through the Fenton-like process, and the generated Fe^{2+} ions are then oxidized immediately by H_2O_2 to complete the Fe^{3+}/Fe^{2+} ions and generate •OH radicals, which is responsible for the dye degradation [23]. Additionally, the photosensitization of the azo dyes can promote a conversion of Fe^{3+} to Fe^{2+} ions during photoassisted Fenton reaction [5, 24]. In this work, it is believed that RR 195 could act as the sensitizer of the Fe-AO-n-PAN or Fe-AO-PAN to lead to a charge transfer with the concomitant quenching of the excited dye* and the formation of dye+. These reaction processes are described by

$$Fe^{3+}/PAN + H_2O_2 \rightarrow Fe^{2+}/PAN + HO_2^\bullet + H$$

$$Fe^{2+}/PAN + H_2O_2 \rightarrow Fe^{3+}/PAN + OH^\theta + {}^\bullet OH$$

$$Fe^{3+}/PAN + HO_2^\bullet \rightarrow Fe^{2+}/PAN + O_2$$

$$Dye + Fe^{3+}/PAN$$

$$\rightarrow Dye^* -- Fe^{3+}/PAN$$

$$\rightarrow \quad Dye^{+\bullet} + Fe^{2+}/PAN$$

$$^\bullet OH + Dye^{+\bullet} \text{ or } Dye \rightarrow Products$$

More importantly, the D% values in the case of Fe-AO-n-PAN are much higher than those in the case of Fe-AO-PAN, especially in the dark at the same conditions. Furthermore, most of RR 195 molecules were completely decolorized in the presence of Fe-AO-n-PAN within 90 min. These results imply that the Fe-AO-n-PAN exhibits a more remarkable activity for the dye degradation than Fe-AO-PAN. This is mainly attributed to the great specific surface area ($3.46 \, m^2 g^{-1}$) of Fe-AO-n-PAN, which is much bigger than that ($0.352 \, m^2 g^{-1}$) of Fe-AO-PAN. Besides, the small diameter of the nanofibers could result in a reduced diffusion resistance of the dye and its degradation products due to the short diffusion path [11, 25], thus providing convenient diffusion channels for dye solution and more reaction sites for dye adsorption and degradation.

11.3.3 FE CONTENT OF CATALYST

The degradation of RR 195 was carried out at pH 6 by the addition of Fe-AO-n-PAN with various Fe content (Q_{Fe}) to the test solutions under light irradiation. The influence of Q_{Fe} value on the D% values of RR 195 within 45 min and 90 min was presented in Figure 5.

It is seen from Figure 5 that the relatively high D% values (40.2% for 45 min and 58.4% for 90 min) were obtained when AO-n-PAN (Q_{Fe} = 0.00 mmol g^{-1}) was used as a catalyst. This is because AO-n-PAN has a big specific surface area, leading to a strong adsorption of the dye molecules in solution. Moreover, increasing values of Fe-AO-n-PAN from 0.801 mmol g^{-1} to 2.25 mmol g^{-1} is accompanied with a significant enhancement of D% levels, which are higher than those in the case of AO-n-PAN. This suggests that higher Q_{Fe} values can significantly accelerate the dye degradation. Similar result was observed for the decomposition of azo dye in the system of Fe-AO-PAN and H_2O_2 in our previous work [6]. The possible reason is that increasing Q_{Fe} values of the catalysts may cause more active sites on their surface, which can improve the H_2O_2 decomposition and generate relatively high concentration of •OH radicals in the solution by the heterogeneous photo-Fenton reaction, thus promoting the dye degradation. It is worth noting that little enhancement was observed at being more than 2.25 mmol g^{-1}, especially for the D_{90}% value. This may be owing to the reaction of excess Fe^{3+} ion with H_2O_2 producing less •OH radicals [26]. Therefore, 2.25 mmol g^{-1} of Q_{Fe} value is believed to be optimum Fe content of Fe-AO-n-PAN for the highest catalytic efficiency.

11.3.4 INITIAL CONCENTRATION OF H2O2

RR 195 was decomposed using Fe-AO-n-PAN (Q_{Fe} = 2.25 mmol g^{-1}) and H_2O_2 with the initial concentration (C_{H2O2}) varied from 0 mmol L^{-1} to 9.00 mmol L^{-1} at pH 6, and the results were shown in Figure 6.

Figure 6 shows that D% value increased with the prolongation of reaction time in the absence or presence of H_2O_2. And D% value reached 35.6% after 90 min without the addition of H_2O_2, which is due to the adsorption effect of Fe-AO-n-PAN with large specific surface area. Moreover, D% values increased with C_{H2O2} increasing from 0 mmol L^{-1} to 3.00 mmol L^{-1}, indicating that high C_{H2O2} levels led to a gradual increasing D% values. However, D% value unremarkably increased when C_{H2O2} is above 3.00 mmol L^{-1}. The optimal C_{H2O2} level was found to be 3.00 mmol L^{-1} in this experiment. This is coincident with the results reported in [27]. According to [28, 29], the role of H_2O_2 in Fenton reaction is different depending on its concentration. At the low C_{H2O2} levels, H_2O_2 molecules could not generate enough •OH radicals, resulting in the limited decoloration of RR 195. When C_{H2O2} was over the optimum value, surplus H_2O_2 molecules may act as •OH scavenger, thus reducing the •OH concentration in solution [27, 30].

11.3.5 IRRADIATION INTENSITY

The degradation of RR 195 was performed at pH 6 with Fe-AO-n-PAN (Q_{Fe} = 2.71 mmol g^{-1}) and 3.0 mmol L^{-1} H_2O_2 under different light irradiations, and D% values within 45 min and 90 min were presented in Table 1.

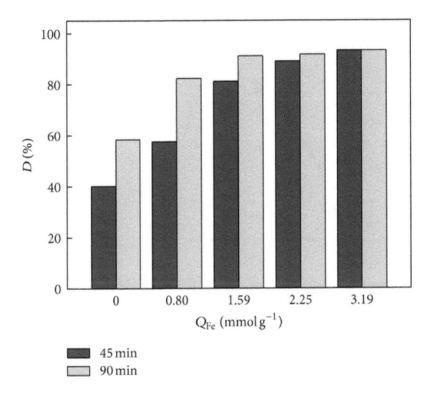

FIGURE 5: Degradation of RR 195 by Fe-AO-n-PAN with different values.

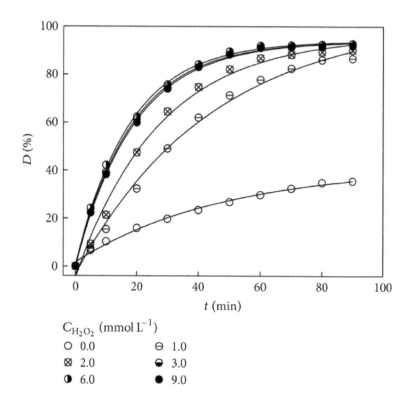

FIGURE 6: Effect of H_2O_2 initial concentration on degradation of RR 195.

TABLE 1: Degradation of RR 195 under increasing light irradiations.

Irradiation intensity	Visible$_{(400-1000\,nm)}$ (m W cm^{-2})	0	8.42	11.3	15.4
	UV$_{(365\,nm)}$ (m W cm^{-2})	0	0.62	0.72	1.04
D%	45 min	61.04	79.47	88.82	98.40
	90 min	82.35	91.25	97.02	99.89

It is noticed from Table 1 that RR 195 molecules were partially removed from the test solution in the dark (no light irradiation), which may be attributed to the better adsorption and catalytic capacity of Fe-AO-n-PAN. Moreover, D% value gradually increased as the irradiation intensity becoming higher. This suggests that stronger light irradiation can significantly enhance catalytic activity of Fe-AO-n-PAN, thus resulting in an obvious improvement in the dye degradation. A main reason is that the amidoximated PAN fiber Fe complex is activated throughout the UV and visible wavelengths where light absorption occurs [6]. Consequently, Fe-AO-n-PAN is regarded as an efficient Fenton photocatalyst for dyeing wastewater. Another explanation is that higher irradiation may increase the photosensitization of the dye molecules on the surface of catalyst, which promotes a conversion of Fe^{3+} to Fe^{2+} ions during the degradation. Besides, higher irradiation may be beneficial for photolysis of H_2O_2 and photoreduction of Fe^{3+} to Fe^{2+} ions in the solution since both processes are strongly dependent on wavelength and intensity of light irradiation [31]. Finally, it is possible that powerful light irradiation may change the aggregation equilibrium of dye molecules in solution, and reduce the formation of dye aggregation units due to their weak stability in aqueous solution [32].

11.3.6 DYE STRUCTURE CHARACTERISTICS AND INITIAL CONCENTRATION

The comparative study was carried out for the degradation of three kinds of textile dyes: azo dye, RR 195; anthraquinone dye, RB 4, and triphenylmethane dye, AB 7 from aqueous solution by Fe-AO-n-PAN (Q_{Fe} = 2.71 mmol g^{-1}) and 3.0 mmol L^{-1} H_2O_2 at pH 6 under light irradiation. The

UV-Vis spectrum analysis was adopted to investigate their degradation and intermediate transformation during treatments. The comparisons of UV-Vis spectra of dye degradation processes are given in Figure 7. Moreover, D% values of three dyes were also calculated and shown in Figure 7.

Figures 7(a)–7(c) show that three characteristic peaks at 522 nm for RR 195, 596 nm for RB 4, and 630 nm for AB 7 decreased gradually with prolonging irradiation time. As shown in Figure 7(d), their D% values were above 90% within 45 min. These results suggest that chromophores of three dyes have been almost completely destructed after the degradation. Not only characteristic peaks of three dyes, but also some peaks (296 nm for RR 195, 256 nm and 296 nm for RB 4, and 307 nm for AB 7) in 200–400 nm also significantly decreased during the degradation processes, which indicated that the aromatic rings of three dye molecules studied were destroyed by the photocatalysis of Fe-AO-n-PAN/H_2O_2. Consequently, Fe-AO-n-PAN was proved to be a universal and efficient catalyst for degradation of the three dyes with different molecular structures.

In order to investigate the effect of dye initial concentration on its degradation, we performed the experiments with four initial concentrations of RR 195 under light irradiation while the other variables were kept constant. It is apparent from Figure 8 that higher initial concentrations can cause a dramatic reduction in D% values, proposing that the lower the RR 195 initial concentration, the shorter the reaction period needed to degrade RR 195 completely. D% values, especially for low RR 195 initial concentration, increased quickly in the first 20 min and then obviously slowed down as time goes on. This is because when the H_2O_2 concentration was kept constant for the solutions with different RR 195 initial concentrations, more H_2O_2 molecules were consumed in the beginning stage [33]. On the other hand, it is well known that the aggregation equilibrium of dye molecules in water is affected particularly by dye concentration [34, 35]. Higher dye concentration can generally enhance the formation of dye aggregation units, which limits the dye adsorption and degradation.

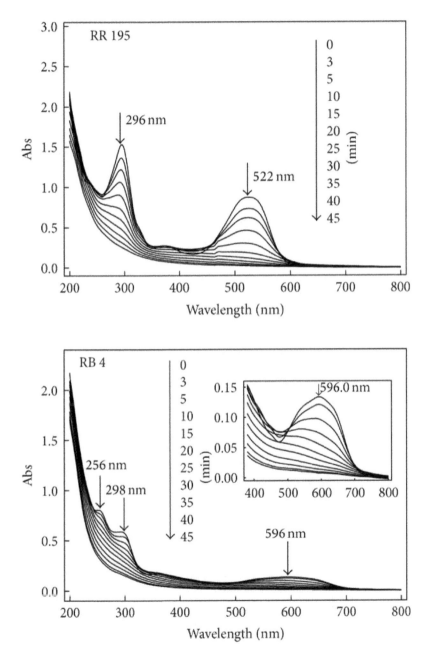

FIGURE 7: Degradation of three dyes with different molecular structures. (a and b)

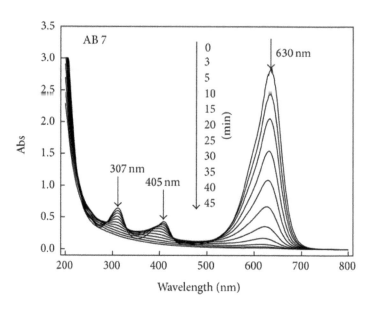

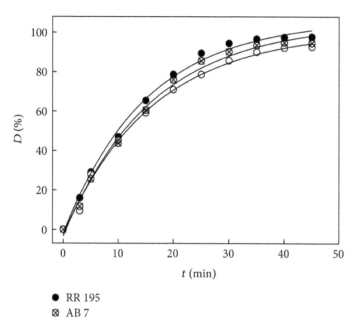

FIGURE 7: *Cont.* (c and d)

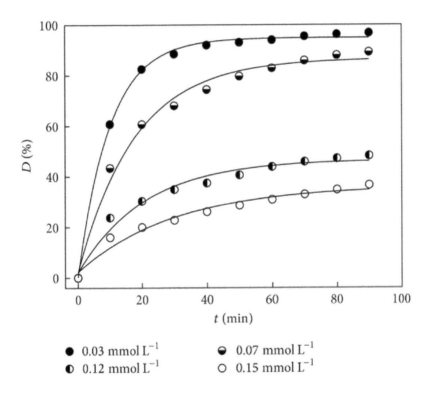

FIGURE 8: The degradation of RR 195 with different initial concentrations, experimental conditions: 0.20 g Fe-AO-n-PAN ($Q_{Fe} = 2.15$ mmolg−1), H_2O_2: 3.0 mmol L^{-1}, and pH = 6.

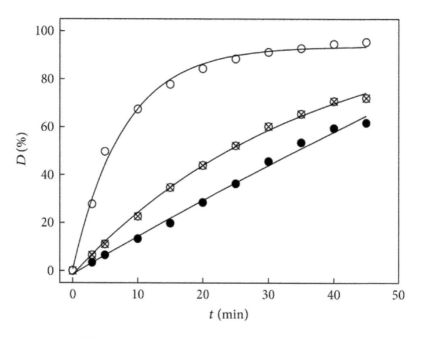

FIGURE 9: Degradation of RR 195 with Fe-AO-n-PAN in different pH levels.

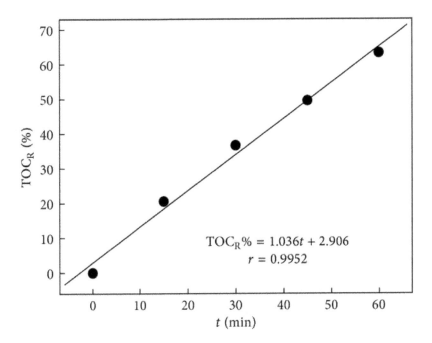

FIGURE 10: $TOC_R\%$ values of RR 195 solution as a function of the reaction time.

11.3.7 THE SOLUTION PH

The degradation of RR 195 with $3.0\,mmol\,L^{-1}$ H_2O_2 catalyzed by Fe-AO-n-PAN ($Q_{Fe} = 3.71\,mmol\,g^{-1}$) at different solution pH was investigated under light irradiation in order to ensure the stability of Fe-AO-n-PAN as a photocatalyst over a wide pH range, and the experimental result was presented in Figure 9.

It was clearly observed from Figure 9 that D% values as a function of the reaction time when initial solution pH is varied. The dye degradation occurred over a wide range of pH values from 3 to 9. D% value reached the maximum value of 95.42% at pH 6 within 45 min. D% value at pH 3 was higher than that at pH 9 at the same reaction time. These results demonstrated that Fe-AO-n-PAN exhibited a higher catalytic activity at acidic medium than at alkaline medium, and its highest catalytic efficiency could be obtained at pH 6. This optimum solution pH is higher than pH level (about 3.0) at which homogeneous photoassisted Fenton reaction was performed most easily [36]. On the other hand, the change in the pH of the reaction system during the degradation of RR 195 was obviously found. For instance, the final pH value after 45 min of reaction was about 2.6 for an initial pH of 6. This may be owing to the acidity of the intermediates produced during the degradation [33]. Some similar phenomena were observed by Feng et al. [33] and Dhananjeyan et al. [37].

11.3.8 MINERALIZATION

Mineralization of the dye molecules in solution should be examined since the degradation intermediates may be longer-lived and even more toxic than the parent dyes to aquatic animals and human beings. In this work, mineralization of RR 195 was performed in the presence of $3.0\,mmol\,L^{-1}$ H_2O_2 and Fe-AO-n-PAN ($Q_{Fe} = 2.71\,mmol\,g^{-1}$) at pH 6 under light irradiation, and the result is given in Figure 10.

Figure 10 shows that TOC_R% value increases proportionally as the reaction time goes on. A linear equation between TOC_R% value and reaction time (t) also is also provided. Moreover, TOC_R% value reaches 63.28% within 60 min under light irradiation. These results illustrates that RR 195 molecules can be destructed and then converted into H_2O, CO_2, and inorganic salts. This is in good agreement with the decoloration and degradation of the RR 195 mentioned above.

11.4 CONCLUSIONS

The modified PAN nanofiber Fe complex was prepared through the amidoximation and consequent Fe coordination of PAN nanofiber obtained using electrospinning technique. This complex not only significantly catalyzed the oxidative degradation of the textile dye in water as a novel heterogeneous Fenton photocatalyst, but also showed a better catalytic performance than the Fe complex prepared with conventional PAN yarns in the dark or under light irradiation. It was found that Fe content of the complex, H_2O_2 initial concentration, irradiation intensity, and the solution pH were the main factors that have strong impacts on the heterogeneous Fenton degradation of the textile dye. Increasing Fe content of the complex or irradiation intensity would accelerate the dye degradation. The optimal H_2O_2 molar concentration is $3.0\,mmol\,L^{-1}$ in order to obtain the best decoloration efficiency. Although this complex has a relatively better catalytic activity over a wide pH range, the optimal pH of the reaction solution is 6 for the fastest degradation of the dye. Moreover, it was proved that the modified PAN nanofiber Fe complex was a universal and efficient catalyst for degradation of the three classes of textile dyes including azo dye, anthraquinone dye, and triphenylmethane dye. Besides, UV-Vis spectrum analysis and TOC measurement indicated that the chromophore and aromatic rings of the dye molecules were decomposed and then converted into H_2O, CO_2, and inorganic salts in the presence of the modified PAN nanofiber Fe complex and H_2O_2 under light irradiation.

REFERENCES

1. B. Lodha and S. Chaudhari, "Optimization of Fenton-biological treatment scheme for the treatment of aqueous dye solutions," Journal of Hazardous Materials, vol. 148, no. 1-2, pp. 459–466, 2007.

2. E. Abadulla, T. Tzanov, S. Costa, K.-H. Robra, A. Cavaco-Paulo, and G. M. Gubitz, "Decolorization and detoxification of textile dyes with a laccase from Trametes hirsuta," Applied and Environmental Microbiology, vol. 66, no. 8, pp. 3357–3362, 2000.

3. A. Mittal, V. Gajbe, and J. Mittal, "Removal and recovery of hazardous triphenylmethane dye, Methyl Violet through adsorption over granulated waste materials," Journal of Hazardous Materials, vol. 150, no. 2, pp. 364–375, 2008.

4. W. J. Epolito, Y. H. Lee, L. A. Bottomley, and S. G. Pavlostathis, "Characterization of the textile anthraquinone dye Reactive Blue 4," Dyes and Pigments, vol. 67, no. 1, pp. 35–46, 2005.

5. Z. Ding, Y. Dong, and B. Li, "Preparation of a modified PTFE fibrous photo-Fenton catalyst and its optimization towards the degradation of organic dye," International Journal of Photoenergy, vol. 2012, Article ID 121239, 8 pages, 2012.

6. Y. Dong, Z. Han, C. Liu, and F. Du, "Preparation and photocatalytic performance of Fe (III)-amidoximated PAN fiber complex for oxidative degradation of azo dye under visible light irradiation," Science of the Total Environment, vol. 408, no. 10, pp. 2245–2253, 2010.

7. L. Wan, Z. Xu, and H. Jiang, "Fibrous membranes electrospinning from acrylonitrue-based polymers: specific absorption behaviors and states of water," Macromolecular Bioscience, vol. 6, no. 5, pp. 364–372, 2006.

8. H. Zhang, H. Nie, D. Yu et al., "Surface modification of electrospun polyacrylonitrile nanofiber towards developing an affinity membrane for bromelain adsorption," Desalination, vol. 256, no. 1–3, pp. 141–147, 2010.

9. K. Yoon, B. S. Hsiao, and B. Chu, "High flux nanofiltration membranes based on interfacially polymerized polyamide barrier layer on polyacrylonitrile nanofibrous scaffolds," Journal of Membrane Science, vol. 326, no. 2, pp. 484–492, 2009.

10. A. Sato, R. Wang, H. Ma, B. S. Hsiao, and B. Chu, "Novel nanofibrous scaffolds for water filtration with bacteria and virus removal capability," Journal of Electron Microscopy, vol. 60, no. 3, pp. 201–209, 2011.

11. Q. Feng, Q. Wang, B. Tang et al., "Immobilization of catalases on amidoxime polyacrylonitrile nanofibrous membranes," Polymer International, vol. 62, no. 2, pp. 251–256, 2013.

12. S. Li, J. Chen, and W. Wu, "Electrospun polyacrylonitrile nanofibrous membranes for lipase immobilization," Journal of Molecular Catalysis B, vol. 47, no. 3-4, pp. 117–124, 2007.

13. S. Li and W. Wu, "Lipase-immobilized electrospun PAN nanofibrous membranes for soybean oil hydrolysis," Biochemical Engineering Journal, vol. 45, no. 1, pp. 48–53, 2009.

14. N. Zhou, T. Yang, C. Jiang, M. Du, and K. Jiao, "Highly sensitive electrochemical impedance spectroscopic detection of DNA hybridization based on Aunano-CNT/PANnano films," Talanta, vol. 77, no. 3, pp. 1021–1026, 2009.

15. P. Kampalanonwat and P. Supaphol, "Preparation and adsorption behavior of aminated electrospun polyacrylonitrile nanofiber mats for heavy metal ion removal," ACS Applied Materials and Interfaces, vol. 2, no. 12, pp. 3619–3627, 2010.

16. K. Saeed, S. Park, and T. Oh, "Preparation of hydrazine-modified polyacrylonitrile nanofibers for the extraction of metal ions from aqueous media," Journal of Applied Polymer Science, vol. 121, no. 2, pp. 869–873, 2011.

17. Q. Feng, X. Wang, A. Wei et al., "Surface modified ployacrylonitrile nanofibers and application for metal ions chelation," Fibers and Polymers, vol. 12, no. 8, pp. 1025–1029, 2011.

18. S. Li, X. Yue, Y. Jing, S. Bai, and Z. Dai, "Fabrication of zonal thiol-functionalized silica nanofibers for removal of heavy metal ions from wastewater," Colloids and Surfaces A, vol. 380, no. 1–3, pp. 229–233, 2011.

19. K. Saeed, S. Haider, T. Oh, and S. Park, "Preparation of amidoxime-modified polyacrylonitrile (PAN-oxime) nanofibers and their applications to metal ions adsorption," Journal of Membrane Science, vol. 322, no. 2, pp. 400–405, 2008.

20. Z. Han, Y. Dong, and S. Dong, "Comparative study on the mechanical and thermal properties of two different modified PAN fibers and their Fe complexes," Materials and Design, vol. 31, no. 6, pp. 2784–2789, 2010.

21. Y. Kim, C. H. Ahn, M. B. Lee, and M. Choi, "Characteristics of electrospun PVDF/SiO2 composite nanofiber membranes as polymer electrolyte," Materials Chemistry and Physics, vol. 127, no. 1-2, pp. 137–142, 2011.

22. Y. Dong, W. Dong, Y. Cao, Z. Han, and Z. Ding, "Preparation and catalytic activity of Fe alginate gel beads for oxidative degradation of azo dyes under visible light irradiation," Catalysis Today, vol. 175, no. 1, pp. 346–355, 2011.

23. Y. Han, Y. Dong, Z. Ding, and C. Liu, "Influence of polypropylene fibers on preparation and performance of Fe-modified PAN/PP blended yarns and their knitted fabrics," Textile Research Journal, vol. 83, no. 3, pp. 219–228, 2013.

24. J. F. Wei, Z. P. Wang, J. Zhang, Y. Y. Wu, Z. P. Zhang, and C. H. Xiong, "The preparation and the application of grafted polytetrafluoroethylene fiber as a cation exchanger for adsorption of heavy metals," Reactive and Functional Polymers, vol. 65, no. 1-2, pp. 127–134, 2005.

25. S. Sakai, K. Antoku, T. Yamaguchi, and K. Kawakami, "Development of electrospun poly(vinyl alcohol) fibers immobilizing lipase highly activated by alkyl-silicate for flow-through reactors," Journal of Membrane Science, vol. 325, no. 1, pp. 454–459, 2008.

26. I. Muthuvel and M. Swaminathan, "Photoassisted Fenton mineralisation of Acid Violet 7 by heterogeneous Fe(III)-Al2O3 catalyst," Catalysis Communications, vol. 8, no. 7, pp. 981–986, 2007.

27. Y. Dong, L. He, and M. Yang, "Solar degradation of two azo dyes by photocatalysis using Fe(III)-oxalate complexes/H2O2 under different weather conditions," Dyes and Pigments, vol. 77, no. 2, pp. 343–350, 2008.

28. S. Kaur and V. Singh, "TiO2 mediated photocatalytic degradation studies of Reactive Red 198 by UV irradiation," Journal of Hazardous Materials, vol. 141, no. 1, pp. 230–236, 2007.

29. C. M. So, M. Y. Cheng, J. C. Yu, and P. K. Wong, "Degradation of azo dye Procion Red MX-5B by photocatalytic oxidation," Chemosphere, vol. 46, no. 6, pp. 905–912, 2002.

30. M. Muruganandham and M. Swaminathan, "Photochemical oxidation of reactive azo dye with UV-H2O2 process," Dyes and Pigments, vol. 62, no. 3, pp. 269–275, 2004.

31. B. Muthukumari, K. Selvam, I. Muthuvel, and M. Swaminathan, "Photoassisted hetero-Fenton mineralisation of azo dyes by Fe(II)-Al2O3 catalyst," Chemical Engineering Journal, vol. 153, no, 1–3, pp 9–15 2009

32. Y. Dong, W. Dong, C. Liu, Y. Chen, and J. Hua, "Photocatalytic decoloration of water-soluble azo dyes by reduction based on bisulfite-mediated borohydride," Catalysis Today, vol. 126, no. 3-4, pp. 456–462, 2007.

33. J. Feng, X. Hu, P. L. Yue, H. Y. Zhu, and G. Q. Lu, "Degradation of azo-dye orange II by a photoassisted Fenton reaction using a novel composite of iron oxide and silicate nanoparticles as a catalyst," Industrial and Engineering Chemistry Research, vol. 42, no. 10, pp. 2058–2066, 2003.

34. W. S. Perkins, Textile Coloration and Finishing, China Textile Press, Beijing, China, 2004.

35. M. M. El-Fass, N. A. Badawy, A. A. El-Bayaa, and N. S. Moursy, "The influence of simple electrolyte on the behavior of some acid dyes in aqueous media," Bulletin of the Korean Chemical Society, vol. 16, no. 5, pp. 458–461, 1995.

36. M. Hartmann, S. Kullmann, and H. Keller, "Wastewater treatment with heterogeneous Fenton-type catalysts based on porous materials," Journal of Materials Chemistry, vol. 20, no. 41, pp. 9002–9017, 2010.

37. M. R. Dhananjeyan, J. Kiwi, P. Albers, and O. Enea, "Photo-assisted immobilized Fenton degradation up to pH 8 of azo dye Orange II mediated by Fe+/Nafion/glass fibers," Helvetica Chimica Acta, vol. 84, no. 11, pp. 3433–3445, 2001.

CHAPTER 12

Ozonation of Indigo Carmine Catalyzed with Fe-Pillared Clay

MIRIAM BERNAL, RUBÍ ROMERO, GABRIELA ROA,
CARLOS BARRERA-DÍAZ, TERESA TORRES-BLANCAS,
AND REYNA NATIVIDAD

12.1 INTRODUCTION

A vast amount of water is employed by the textile industry. Because of the dyeing process, the produced wastewater contains strong color and this is reflected as well in a high chemical oxygen demand (COD). It has been estimated that 1–15% of the dye is lost during dyeing and finishing processes and is released into wastewater [1, 2]. Even small quantities ($<0.005\,mg\,L^{-1}$) of dyes in water are unacceptable since the discharge of effluents containing reactive dyes into the environment can interfere with transmission of sunlight into flowing liquid [3]. This causes perturbations in aquatic life and the food web [4]. Thus, an effective and economical technique for removing dyes from textile wastewaters is needed [5]. In this sense, several conventional methods for treating dye effluents have been studied, such as photodegradation [6], adsorption [7], filtra-

Ozonation of Indigo Carmine Catalyzed with Fe-Pillared Clay. © *Bernal M, Romero R, Roa G, Barrera-Díaz C, Torres-Blancas T, and Natividad R.* International Journal of Photoenergy **2013** (2013). *ttp://dx.doi.org/10.1155/2013/918025. Licensed under Creative Commons Attribution 3.0 Unported License, http://creativecommons.org/licenses/by/3.0/.*

tion [8], coagulation [9], and biological treatments [10]. However, due to the stability of the molecules of dyes some of these methods are not completely effective and/or viable. In recent years, water treatment based on the chemical oxidation of organic compounds by advanced oxidation processes (AOPs) like ozonation has drawn attention. Ozonation, which is effective, versatile, and environmentally sound, has been tested as a good method for color removal [11]. Ozone is a strong oxidant ($E°=2.07$ V) and reacts rapidly with most of organic compounds [12]. It oxidizes organic pollutants via two pathways: direct oxidation with ozone molecules and/ or the generation of free-radical intermediates, such as the •OH radical, which is a powerful, effective, and nonselective oxidizing agent [13, 14]. However, the degree and rate of oxidation by ozonation is limited by the chemical structure; that is, amines, linear chain alcohols, and ketones are harder to be oxidized by ozone than some aromatic compounds [15]. This has motivated the search for more efficient ozonation processes. In this context, metal-catalyzed ozonations have particularly drawn attention, but water contamination by metals turned out to be a major problem [16, 17]. The aim of this work is to evaluate the efficiency of ozonation of carmine indigo catalyzed by an Fe-pillared clay. The clay is expected to retain the iron within its structure so that the solution is not contaminated with the metal.

Clays are natural abundant minerals that can be obtained from some mines in high purity [18]. However, it is important to modify them to obtain homogeneous materials that can be used with reliable results. Furthermore, although clay possesses large surface area, this is not accessible due to the strong electrostatic interaction between sheets and charge balancing cations. Thus, to gain access to the interlaminar area large cations ("pillars") should be placed between sheets; as a result an increase in the surface area is obtained. Pillared clays are microporous materials that are obtained by exchanging the interlayered cations of layered clays with bulky inorganic polyoxocations, followed by calcinations [19–22]. After pillaring, the presence of this new porous structure and the incorporation of new active sites present several possible applications of these materials [23]. As described by Catrinescu et al. [24], there are few examples where iron containing synthetic clays have been tested as solid catalyst to promote the Fenton reaction [24]. Some reports indicate that pillared clays can be used as Fenton catalyst; these reports refer to the catalytic activity imparted by the pillars, which become the actual catalytic sites [25–27]. However, in all the previous research the addition of hydrogen peroxide has been the common reactive added to promote the Fenton reaction. To the authors knowledge there are no reports of the efficiency of ozonation of dyes catalyzed with Fe-pillared clays.

12.2 MATERIALS AND METHODS

12.2.1 REAGENTS

Sulfuric acid, sodium hydroxide, and indigo carmine dye of analytical grade were purchased from Sigma-Aldrich Chemicals. The acid and the base were used without further purification to adjust the solutions pH to 3. Purified-grade bentonite was supplied by Fisher Scientific.

Ozone was generated in situ from dry air by an ozone generator (Pacific Ozone Technology), with an average ozone production of $0.005 \, g \cdot dm^{-3}$.

12.2.2 SYNTHESIS OF FE-PILCS

The Fe-pillared clay (Fe-PILC), used as catalyst, was prepared by a purified-grade bentonite supplied by Fisher Scientific. Fe-PILCs were prepared using $FeCl_3 \cdot 6H_2O$ and NaOH solutions according to the method of synthesis reported elsewhere [28]. The former was added to the NaOH solution to obtain the required OH/Fe molar ratio of 2.0. The initial concentration of the Fe salt was 0.2 M and the suspension clay concentration was 0.10 wt.%. In order to avoid precipitation of iron species, the pH was kept constant at 1.78–1.80. The mixture was aged for 4 h under stirring at room temperature. The pillaring solution was then slowly added to a suspension of bentonite in deionized water. The mixture was kept under vigorous stirring for 12 h at room temperature. Finally, the solid was washed by vacuum filtration with deionized water until it was chloride-free (conductivity < 10 µS/cm). Finally, the solid was air-dried (70°C) and calcined for 2 h at 400°C.

12.2.3 CATALYST CHARACTERIZATION

X-ray diffraction (XRD) patterns were obtained on a Bruker Advance 8 diffractometer using CuKα radiation at 35 kV and 30 mA. Data were collected over 2θ range of 3–12° with a step of 0.04°/min. X-ray diffraction pattern suggests that the original bentonite was successfully pillared since one peak corresponding to the (0 0 1) reflection appeared at small 2θ angles (2θ ≈ 4°). According to [28] this result clearly indicates an enlargement of the basal spacing of the clay as consequence of the pillaring process.

The total iron content incorporated into the catalyst was determined by using a SpectrAA 240FS atomic absorption spectrophotometer. Before analysis, the samples were dissolved in hydrofluoric acid and diluted to the interval of measurement.

Specific surface area and pore-size distribution were determined by N_2 adsorption at 77 K in a static volumetric apparatus (Micromeritics ASAP 2010 sorptometer). Pillared clays were outgassed prior use at 180°C for 16 h under vacuum of 6.6×10^{-9} bar. Specific total surface area was calculated using the Brunauer-Emmett-Teller (BET) equation, whereas specific total pore volume was evaluated from nitrogen uptake at N_2 relative pressure of $P/P_0 = 0.99$.

The chemical state of the Fe was analyzed by X-ray photoelectron spectroscopy (XPS, Jeol JPS 9200) with a standard Mg Kα excitation source (1253.6 eV). Binding energies were calibrated with respect to the carbon signal (C1s) at 285 eV.

12.2.4 OZONATION EXPERIMENTS

The ozonation experiments were conducted in an up-flow glass bubble column reactor (Figure 1). The gas mixture ozone/air was continuously fed with a flow rate varying from 0.02 to 0.06 L·min⁻¹ through a gas diffuser with a 2 mm pore size at the bottom of the reactor. The excess of ozone in the outlet gas was decomposed and trapped in a KI solution. Samples were taken at specific time intervals to be analyzed. Fe-pillared clays were crushed and sieved to a particle size of 60 μm. All the experiments were carried out at room temperature (19°C ± 2). pH was adjusted at 3.0 with analytical grade sulfuric acid and sodium hydroxide. In addition to ozone flow rate, the effect of mass of catalyst was also studied in the range of 0–0.1% w/w. At all experiments the initial concentration of IC was 1000 mg/L. As control experiments, IC concentration profiles were established by adsorption, ozonation plus bentonite (clay without pillaring) and particles of Fe°. For the adsorption experiment, only clay without any ozone supplying was employed in order to discard the removal of IC by physical means. The experiment with only bentonite was conducted with an ozone flow rate of $0.045 \, L \cdot min^{-1}$ and a bentonite loading of 0.1%. It is worth clarifying that in this case the employed bentonite was not pillared. For the experiment with Fe° particles, a loading of 0.016% w/w was employed. This iron loading is equivalent to that when using 0.1% w/w of Fe-pillared clay. Fe° particles were synthesized in our laboratory by reducing 0.01 M Fe(II) sulfate solution with $NaBH_4$ at pH 7 and room temperature. pH was controlled by adding a 0.5 M NaOH solution. This was carried out under nitrogen atmosphere and with deaerated water to ensure the production of Fe° particles.

12.2.5 CHEMICAL ANALYSIS

Concentration of indigo carmine was determined by UV-Vis spectrophotometry technique, using a Perkin-Elmer Model Lambda 25 UV-Vis spectrophotometer with a wavelength range of 190–1100 nm. Samples absorbance was scanned from 200 to 900 nm, and a maximum absorbance at 611 nm was observed. The scan rate was $960 \, nm \cdot s^{-1}$. The samples were scanned in a quartz cell with 1 cm optical path.

In order to establish the oxidation degree of indigo carmine, chemical oxygen demand of samples was determined by means of the American Public Health Association (APHA) standard procedures [29].

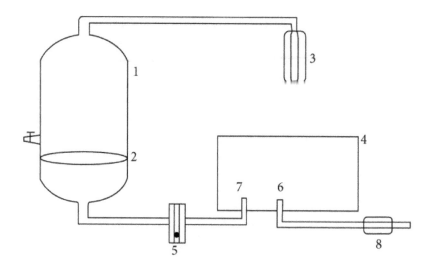

FIGURE 1: Schematic apparatus for the ozonation reaction. (1) Up-flow glass bubble column reactor, (2) porous glass (gas diffuser), (3) KI solution trap, (4) ozone generator, (5) flow meter, (6) dry air inlet, (7) ozone output, and (8) air dryer.

12.3 RESULTS AND DISCUSSION

12.3.1 FE-PILCS CHARACTERIZATION

The XRD pattern of Fe-PILC after calcinations at 400°C is shown in Figure 2. This pillared material exhibits a peak at $2\theta \approx 4°$ which is commonly assigned to the basal (0 0 1) reflection (d(0 0 1)). The basal spacing represents the distance between two clay layers, including the thickness of one of the layers. The reflection at low 2θ values ($2\theta \approx 4°$) is a clear indication of the enlargement of the basal spacing of the clay as consequence of the pillaring process, as explained in a previous work [28]. In this pillared clay a broader peak in the range $2\theta \approx 7$–$9°$ was observed. This can be ascribed to two overlapped peaks, the (0 0 2) reflection of the pillared clay and the (0 0 1) reflection corresponding to the intercalation of a proportion of monomeric species of small size, thus leading to smaller openings of the clay layers [30].

Table 1 shows the main textural characteristics, the Fe content of the samples prepared, and the basal spacing corresponding to both the parent clay and the Fe-PILC calcined at 400°C. As can be seen, the pillared process increases the basal spacing of the clay. The surface area analysis indicates that the pillaring process produces a significant increase in the surface area of parent clay. This can be ascribed to the micropores formation [31].

TABLE 1: Main characteristics of parent clay and Fe-PILC catalyst.

Sample	d (0 0 1) (Å)	Surface area (m²/g)	Micropore volume (cm³/g)	Fe (wt.%)
Bentonite	—, 9.6[b]	35	0.0028	2.7
Fe-PILC	24.22[a], 11.6[b]	283	0.125	16

[a] *1st peak (2θ ≈ 4°).*
[b] *2nd peak (2θ ≈ 7–9°).*

XPS analysis was performed in order to corroborate the presence of iron and to identify its chemical state. The XPS spectrum corresponding to the Fe $2p^{3/2}$ narrow scan region of the Fe-PILC was observed (data not shown). According to the National Institute of Standards and Technology (NIST), this peak (≈710 eV) corresponds to FeO (Fe^{2+}). This result is coherent with TPR studies that indicated that Fe-pillared clays present only one reduction peak that corresponds to the $Fe^{3+} \rightarrow Fe^{2+}$ reduction process [28].

12.3.2 OZONATION EXPERIMENTS

12.3.2.1 EFFECT OF OZONE FLOW RATE

In an up-flow bubble column the gas flow rate is expected to affect, among other aspects, the rising bubble velocity and therefore the contact time of the gas with the solution. This is finally reflected in the gas mass transferred to solution. Thus, gas flow rate is an important variable in the study of any three-phase system. In this case, to do so, this variable was studied only under the presence of ozone in the range of 0.020–0.060 L/min. Figure 3 shows the obtained averaged results of two repetitions per flow rate. It can be observed that at all experiments 100% degradation of indigo carmine (IC) was achieved. The velocity at which this occurs, however, depends on ozone flow rate. It can be observed that depletion of indigo carmine concentration along time increases when increasing of flow rate. Nevertheless, when flow rate is $0.06 L \cdot min^{-1}$ the rate of indigo carmine degradation becomes slower. This means that a maximum of ozone in solution is achieved before this flow rate, and this is observed to happen when a $0.045 L \cdot min^{-1}$ is employed. Therefore, the rest of the experiments were conducted at this flow rate.

12.3.2.2 EFFECT OF MASS OF CATALYST

In order to study the effect of this variable four experiments with their corresponding repetitions were conducted within the catalyst concentration range of 0–0.1% w/w and with an initial concentration of IC at all cases of 1000 mg/L and an ozone flow rate of $0.045 L \cdot min^{-1}$. All the experiments were carried out at pH 3 to avoid the precipitation of Fe [32]. From Figures 4 and 5, the positive effect of the Fe-pillared clay is evident. The presence of the catalyst enhances the ozonation process, and the mass of catalyst has a strong influence over the dye degradation. This increases when mass of Fe-pillared clays increases and this is also an indication of the liquid-solid mass transfer resistance being negligible. It is observed in Figure 4 that by only adding 0.1% w/w of catalyst the IC degradation rate is doubled. Moreover, Figure 5 shows that the use of catalyst not only affects IC degradation rate but also its oxidation degree. Three other plots can be observed in Figure 4. The adsorption experiment confirmed the role of Fe-PILC as catalyst rather than as sorbent. It was found that only 2.9% of IC is removed by this phenomenon during the time of reaction (60 min). Also, in Figure 4, the experiment labeled as bentonite further highlights the effect of pillaring since it can be observed that the bentonite alone does not have a significant effect on ozonation rate, which came out to be practically the same when using only ozone. It is worth clarifying that in this case the employed bentonite was not pillared and that its iron content (by nature) was determined by atomic absorption to be 2.7%. Regarding the effect of Fe particles, it can be observed that the initial IC degradation rate is noticeably improved by the presence of this solid. This improvement, however, is not constant and the degradation rate becomes relatively slower after few minutes of reaction. This behavior can be ascribed to the ready availability of Fe for reactions (4) and (5) to occur. It seems, however, that in this case, reaction (4) occurs more rapidly than reaction (6), which implies that after few minutes of reaction Fe^{3+} species may prevail thus limiting degradation rate.

Figure 5 indicates that the oxidation degree depends on catalyst concentration. To elucidate the species that became resistant to each treatment the UV-Vis spectra of the reacting solution as function of time were analyzed and these are depicted in Figure 6. It can be observed in this figure that the corresponding spectrum to IC presents two absorption bands with maxima at 600 and 340 nm. The former band is characteristic of indigo and the latter is ascribed to auxochromes (N, SO_3) joined to the benzene ring. When ozone is applied to the IC solution, it can be observed in Figure 6(a) that the maximum absorbance at 600 nm decreases and this indicates that the characteristic blue color is diminishing also. This may be due to the loss of sulphonate group since this works as auxochrome and therefore increases color intensity. After 40 minutes of reaction a new absorption band appears at 403 nm and this can be ascribed to a degradation product of IC. This product was identified as isatine by comparison with the absorption spectra of the corresponding standard. After 120 minutes of ozonation treatment only isatine is identified in the solution and no further degradation with ozone only was detected.

Unlike with ozonation alone, it can be observed in Figure 6(b) that when using ozonation + Fe PILC not only the loss of color is improved but also the degradation of isatine. Until minute 30, both treatments are very similar. When using Fe-PILC, the absorption band attributed to the IC completely disappears and then the band related to isatine is observed. This diminished up to 72% of that at 60 minutes.

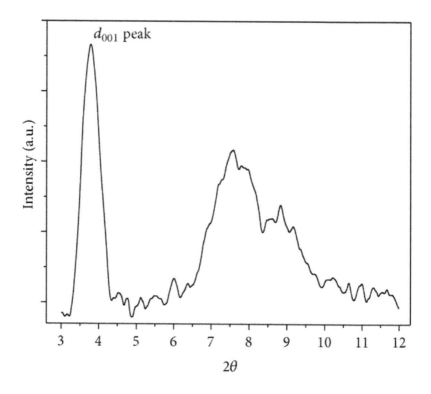

FIGURE 2: X-ray diffraction pattern of Fe-PILC.

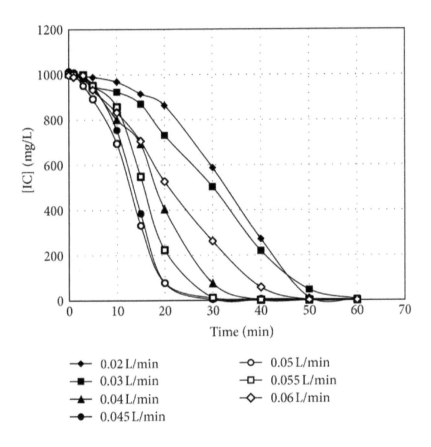

FIGURE 3: Influence of gas flow rate on the indigo carmine concentration.

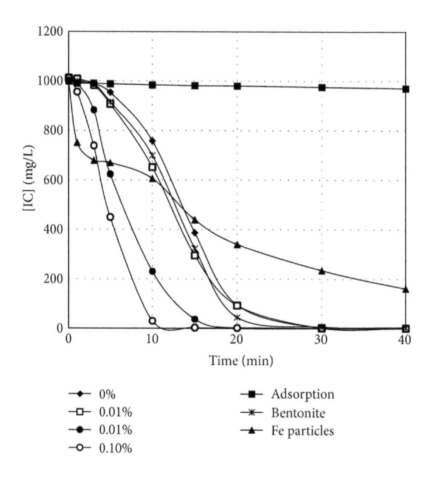

FIGURE 4: Effect of Fe-PILC concentration (% w/w) and type of catalyst (Fe-PILC, clay without pillaring, and Fe particles) on Indigo carmine concentration.

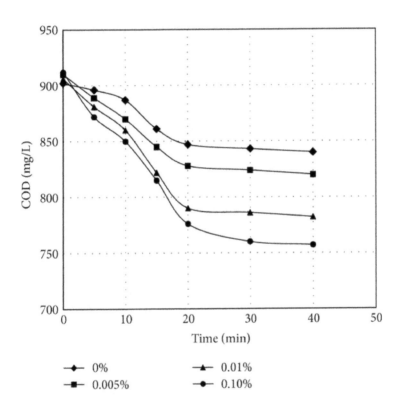

FIGURE 5: Effect of catalyst concentration on chemical oxygen demand (COD).

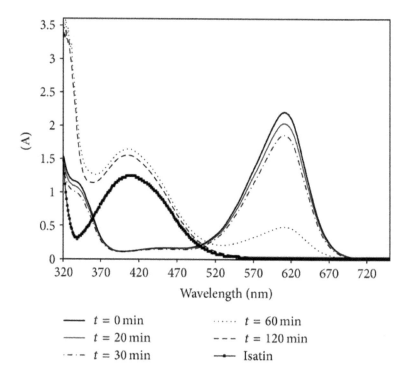

FIGURE 6: Effect of degradation type and time on UV-Vis spectra. (a) Ozonation alone; (b) ozonation + Fe-pillared clay

These results suggest that when only ozone is used the dye is degraded mainly by the direct action of ozone molecules. This, however, does not rule out the production of •OH radicals by means of reaction (1). These radicals may be consumed by reaction (2) rather than by oxidizing the IC. The product of this reaction may be reacting further to obtain H_2O_2 [33]:

$$O_3 + H_2O \rightarrow 2OH^\bullet + O_2 \tag{1}$$

$$O_3 + OH^\bullet \rightarrow O_2 + HO_2^\bullet \leftrightarrows O_2^\bullet + H \tag{2}$$

$$2OH_2^\bullet \rightarrow O_2 + H_2O_2 \tag{3}$$

When the ozone is combined with Fe-PILC the intensification of OH production is expected according the following reaction [34]:

$$FeO^{2+} + H_2O_2 \rightarrow Fe^{3+} + OH^\bullet + OH \tag{4}$$

The ferrous species involved in reaction (4) may be produced by the ferrous species in the pillared clay by means of the following reactions:

$$6H^+ + O_3 + FeO \rightarrow FeO^{2+} + 3H_2O \tag{5}$$

Fe^{3+} species may be going back to Fe^{2+} by means of the following reaction:

$$Fe^{3+} + H_2O_2 \rightarrow Fe^{2+} + HO_2^\bullet + H \tag{6}$$

Therefore, when adding the Fe-PILC to the ozonation process, the oxidation reactions are intensified and so the degradation of IC to isatine and other products is plausible to occur according to Scheme 1 [35].

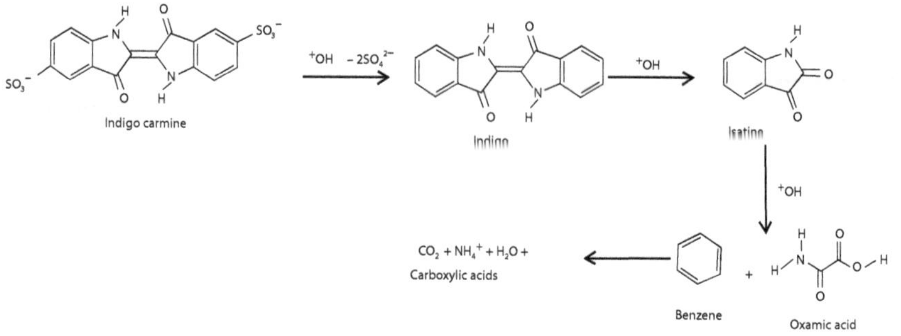

SCHEME 1: Plausible mechanism of degradation of indigo carmine by ozonation catalyzed with Fe-pillared clay.

Figure 7 shows the evolution of pH with time. It is observed that pH slightly rises at the beginning of the reaction and this may be ascribed to the expected production of HO radicals. After the initiation period, pH drops down to 2 and this may be due to the acid products shown in scheme of reaction (1).

XPS analysis was made to determine the oxidation state of the Fe present in the clay at the end of reaction and it was demonstrated to be 2+. It is worth noticing that iron leaching was not observed at any point of reaction.

12.4 CONCLUSIONS

The degradation of indigo carmine by ozonation and catalyzed ozonation with Fe-pillared clay was studied. The role of the Fe-pillared clay as catalyst was demonstrated. A small amount of the clay allows substantially enhancing the degradation and mineralization of indigo carmine. In the catalyzed ozonation process the reaction occurs faster than in the noncatalyzed system doubling the reaction rate. The reutilization of the catalyst is feasible since neither leaching or Fe oxidation state changes were detected.

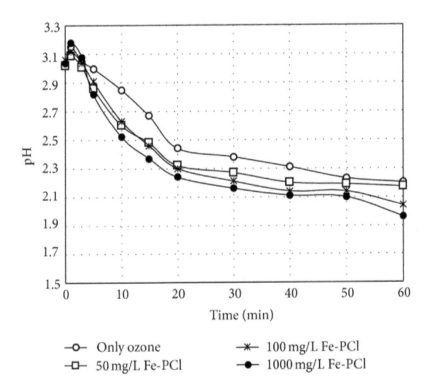

FIGURE 7: pH evolution with time.

REFERENCES

1. N. Daneshvar, M. Rabbani, N. Modirshahla, and M. A. Behnajady, "Photooxidative degradation of Acid Red 27 in a tubular continuous-flow photoreactor: influence of operational parameters and mineralization products," Journal of Hazardous Materials, vol. 118, no. 1–3, pp. 155–160, 2005.
2. N. Daneshvar, H. Ashassi-Sorkhabi, and A. Tizpar, "Decolorization of orange II by electrocoagulation method," Separation and Purification Technology, vol. 31, no. 2, pp. 153–162, 2003.
3. K. Santhy and P. Selvapathy, "Removal of reactive dyes from wastewater by adsorption on coir pith activated carbon," Bioresource Technology, vol. 97, no. 11, pp. 1329–1336, 2006.
4. Z. Aksu, "Reactive dye bioaccumulation by Saccharomyces cerevisiae," Process Biochemistry, vol. 38, no. 10, pp. 1437–1444, 2003.
5. B. Armagan, O. Ozdemir, M. Turan, and C. Elik, "Adsorption of negatively charged azo dyes onto surfactant-modified sepiolite," Journal of Environmental Engineering, vol. 129, no. 8, pp. 709–715, 2003.
6. P. Esparza, M. E. Borges, L. Díaz, and M. C. Alvarez-Galván, "Photodegradation of dye pollutants using new nanostructured titania supported on volcanic ashes," Applied Catalysis A, vol. 388, no. 1-2, pp. 7–14, 2010.
7. R. Gong, Y. Ding, M. Li, C. Yang, H. Liu, and Y. Sun, "Utilization of powdered peanut hull as biosorbent for removal of anionic dyes from aqueous solution," Dyes and Pigments, vol. 64, no. 3, pp. 187–192, 2005.
8. F. I. Hai, K. Yamamoto, F. Nakajima, and K. Fukushi, "Removal of structurally different dyes in submerged membrane fungi reactor—Biosorption/PAC-adsorption, membrane retention and biodegradation," Journal of Membrane Science, vol. 325, no. 1, pp. 395–403, 2008.
9. J.-W. Lee, S.-W. Choi, R. Thiruvenkatachari, W.-G. Shim, and H. Moon, "Evaluation of the performance of adsorption and coagulation processes for the maximum removal of reactive dyes," Dyes and Pigments, vol. 69, no. 3, pp. 196–203, 2006.
10. N. P. Tantak and S. Chaudhari, "Degradation of azo dyes by sequential Fenton's oxidation and aerobic biological treatment," Journal of Hazardous Materials, vol. 136, no. 3, pp. 698–705, 2006.
11. S. Song, J. Yao, Z. He, J. Qiu, and J. Chen, "Effect of operational parameters on the decolorization of C.I. Reactive Blue 19 in aqueous solution by ozone-enhanced electrocoagulation," Journal of Hazardous Materials, vol. 152, no. 1, pp. 204–210, 2008.
12. S. M. D. A. G. Ulson de Souza, K. A. Santos Bonilla, and A. A. Ulson de Souza, "Removal of COD and color from hydrolyzed textile azo dye by combined ozonation and biological treatment," Journal of Hazardous Materials, vol. 179, no. 1–3, pp. 35–42, 2010.
13. B. Merzouk, B. Gourich, K. Madani, C. Vial, and A. Sekki, "Removal of a disperse red dye from synthetic wastewater by chemical coagulation and continuous electrocoagulation. A comparative study," Desalination, vol. 272, no. 1–3, pp. 246–253, 2011.

14. E. Guinea, E. Brillas, F. Centellas, P. Cañizares, M. A. Rodrigo, and C. Saez, "Oxidation of enrofloxacin with conductive-diamond electrochemical oxidation, ozonation and Fenton oxidation. A comparison," Water Research, vol. 43, no. 8, pp. 2131–2138, 2009.

15. L. A. Bernal-Martínez, C. Barrera-Díaz, C. Solís-Morelos, and R. Natividad, "Synergy of electrochemical and ozonation processes in industrial wastewater treatment," Chemical Engineering Journal, vol. 165, no. 1, pp. 71–77, 2010.

16. M. Koch, A. Yediler, D. Lienert, G. Insel, and A. Kettrup, "Ozonation of hydrolyzed azo dye reactive yellow 84 (CI)," Chemosphere, vol. 46, no. 1, pp. 109–113, 2002.

17. K. Pachhade, S. Sandhya, and K. Swaminathan, "Ozonation of reactive dye, Procion red MX-5B catalyzed by metal ions," Journal of Hazardous Materials, vol. 167, no. 1–3, pp. 313–318, 2009.

18. S. Navalon, M. Alvaro, and H. Garcia, "Heterogeneous Fenton catalysts based on clays, silicas and zeolites," Applied Catalysis B, vol. 99, no. 1-2, pp. 1–26, 2010.

19. V. Rives and M. A. Ulibarri, "Layered double hydroxides (LDH) intercalated with metal coordination compounds and oxometalates," Coordination Chemistry Reviews, vol. 181, no. 1, pp. 61–120, 1999.

20. A. Czímerová, J. Bujdák, and R. Dohrmann, "Traditional and novel methods for estimating the layer charge of smectites," Applied Clay Science, vol. 34, no. 1–4, pp. 2–13, 2006.

21. E. M. Serwicka and K. Bahranowski, "Environmental catalysis by tailored materials derived from layered minerals," Catalysis Today, vol. 90, no. 1-2, pp. 85–92, 2004.

22. A. Vaccari, "Preparation and catalytic properties of cationic and anionic clays," Catalysis Today, vol. 41, no. 1–3, pp. 53–71, 1998.

23. A. Gil, L. M. Gandía, and M. A. Vicente, "Recent advances in the synthesis and catalytic applications of pillared clays," Catalysis Reviews, vol. 42, no. 1-2, pp. 145–212, 2000.

24. C. Catrinescu, C. Teodosiu, M. Macoveanu, J. Miehe-Brendlé, and R. le Dred, "Catalytic wet peroxide oxidation of phenol over Fe-exchanged pillared beidellite," Water Research, vol. 37, no. 5, pp. 1154–1160, 2003.

25. S. Perathoner and G. Centi, "Wet hydrogen peroxide catalytic oxidation (WHPCO) of organic waste in agro-food and industrial streams," Topics in Catalysis, vol. 33, no. 1–4, pp. 207–224, 2005.

26. L. F. Liotta, M. Gruttadauria, G. D. Carlo, G. Perrini, and V. Librando, "Heterogeneous catalytic degradation of phenolic substrates: catalysts activity," Journal of Hazardous Materials, vol. 162, no. 2-3, pp. 588–606, 2009.

27. G. Centi and S. Perathoner, "Catalysis by layered materials: a review," Microporous and Mesoporous Materials, vol. 107, no. 1-2, pp. 3–15, 2008.

28. J. L. Valverde, A. Romero, R. Romero, P. B. García, M. L. Sánchez, and I. Asencio, "Preparation and characterization of Fe-PILCS. Influence of the synthesis parameters," Clays and Clay Minerals, vol. 53, no. 6, pp. 613–621, 2005.

29. American Public Health Association American Water Works Association/Water Environment Federation, Standard Methods for the Examination of Water and Wastewater, American Public Health Association American Water Works Association/Water Environment Federation, Washington, DC, USA, 20th edition, 1998.

30. L. S. Cheng, R. T. Yang, and N. Chen, "Iron oxide and chromia supported on titania-pillared clay for selective catalytic reduction of nitric oxide with ammonia," Journal of Catalysis, vol. 164, no. 1, pp. 70–81, 1996.

31. C. Pesquera, F. Gonzalez, I. Benito, S. Mendioroz, and J. A. Pajares, "Synthesis and characterization of pillared montmorillonite catalysts," Applied Catalysis, vol. 69, no. 1, pp. 97–104, 1991.

32. D. Zhao, C. Ding, C. Wu, and X. Xu, "Kinetics of ultrasound-enhanced oxidation of p-nitrophenol by fenton's reagent," Energy Procedia A, vol. 16, pp. 146–152, 2012.

33. M. Hernández-Ortega, T. Ponziak, C. Barrera-Díaz, M. A. Rodrigo, G. Roa-Morales, and B. Bilyeu, "Use of a combined electrocoagulation-ozone process as a pre-treatment for industrial wastewater," Desalination, vol. 250, no. 1, pp. 144–149, 2010.

34. T. Loegager, J. Holcman, K. Sehested, and T. Pedersen, "Oxidation of ferrous ions by ozone in acidic solutions," Inorganic Chemistry, vol. 31, no. 17, pp. 3523–3529, 1992.

35. C. Flox, S. Ammar, C. Arias, E. Brillas, A. V. Vargas-Zavala, and R. Abdelhedi, "Electro-Fenton and photoelectro-Fenton degradation of indigo carmine in acidic aqueous medium," Applied Catalysis B, vol. 67, no. 1-2, pp. 93–104, 2006.

Author Notes

CHAPTER 1

Acknowledgement
This study was supported by grants from the National University of La Plata (Grant 11/N699) and the National Council for Scientific and Technological Research (CONICET, PIP N° 0344) from Argentina.

CHAPTER 2

Acknowledgement
This work was supported by the Spanish MCI through the Project CTM2010-15682.

CHAPTER 3

Author Contributions
Conceived and designed the experiments: HL YW LY. Performed the experiments: HL YW. Analyzed the data: HL YW. Contributed reagents/materials/analysis tools: SZ LY. Wrote the paper: HL SZ YW. Preparation of the manuscript: HL. Finance support and valuable advice: LY. Improvement of manuscript English: SZ.

CHAPTER 4

Acknowledgment
This study was supported by the Research Funds of Jadavpur University, Jadavpur, Kolkata, India.

CHAPTER 7

Acknowledgments

This work was supported by Comunidad de Madrid-Universidad Autóno ma de Madrid through the project CCG06-UAM/AMB-0048 and by Spanish MCI through the projects CTQ2008-03988 and CTM2010-15682.

CHAPTER 9

Acknowledgments

The authors are grateful to CICYT Project CTQ2011-26258, Consolider-Ingenio NOVEDAR 2010 CSD2007-00055, and AGAUR, Generalitat de Catalunya (Project 200956R 1466) for funds received to carry out this work.

CHAPTER 10

Competing Interests

The authors declare that they have no competing interests.

Author Contributions

AL performed the main part of the experiments and drafted the manuscript. MB performed ozone treatment experiments and helped analyzing the data. CG and SS helped interpreting the results and coordinated the manuscript writing. BB helped analyzing the data and drafting the manuscript. All the authors read and approved the final manuscript.

Acknowledgments

The authors gratefully acknowledge the city of Repentigny for providing sewage and biosolid samples. This work was funded by the Chemical Management Plan – Health Canada and the St. Lawrence Action Plan.

CHAPTER 11

Acknowledgments
The authors thank the Tianjin Municipal Science and Technology Committee for the Research Programs of Application Foundation and Advanced Technology (11JCZDJ24600, 11ZCKFGX03200). This research was also supported in part by the grants from the Natural Science Foundation of China (20773093, 51102178).

CHAPTER 12

Conflict of Interest
The authors declare that there is no conflict of interests regarding the publication of this paper.

Acknowledgments
The authors are grateful to CONACYT for financial support of Project 168305. Bernal is grateful to CONACYT for scholarship.

Index

Milton Keynes UK
Ingram Content Group UK Ltd.
UKHW050257161024
449569UK00042B/1753